Martin Romauch

Facility Location and related Problems

Martin Romauch

Facility Location and related Problems

Extensions, Applications, Algorithms and Complexity Issues

Südwestdeutscher Verlag für Hochschulschriften

Impressum / Imprint
Bibliografische Information der Deutschen Nationalbibliothek: Die Deutsche Nationalbibliothek verzeichnet diese Publikation in der Deutschen Nationalbibliografie; detaillierte bibliografische Daten sind im Internet über http://dnb.d-nb.de abrufbar.
Alle in diesem Buch genannten Marken und Produktnamen unterliegen warenzeichen-, marken- oder patentrechtlichem Schutz bzw. sind Warenzeichen oder eingetragene Warenzeichen der jeweiligen Inhaber. Die Wiedergabe von Marken, Produktnamen, Gebrauchsnamen, Handelsnamen, Warenbezeichnungen u.s.w. in diesem Werk berechtigt auch ohne besondere Kennzeichnung nicht zu der Annahme, dass solche Namen im Sinne der Warenzeichen- und Markenschutzgesetzgebung als frei zu betrachten wären und daher von jedermann benutzt werden dürften.

Bibliographic information published by the Deutsche Nationalbibliothek: The Deutsche Nationalbibliothek lists this publication in the Deutsche Nationalbibliografie; detailed bibliographic data are available in the Internet at http://dnb.d-nb.de.
Any brand names and product names mentioned in this book are subject to trademark, brand or patent protection and are trademarks or registered trademarks of their respective holders. The use of brand names, product names, common names, trade names, product descriptions etc. even without a particular marking in this work is in no way to be construed to mean that such names may be regarded as unrestricted in respect of trademark and brand protection legislation and could thus be used by anyone.

Verlag / Publisher:
Südwestdeutscher Verlag für Hochschulschriften
ist ein Imprint der / is a trademark of
OmniScriptum GmbH & Co. KG
Heinrich-Böcking-Str. 6-8, 66121 Saarbrücken, Deutschland / Germany
Email: info@svh-verlag.de

Herstellung: siehe letzte Seite /
Printed at: see last page
ISBN: 978-3-8381-0208-5

Zugl. / Approved by: Wien, Universität Wien, Diss., 2008

Copyright © 2008 OmniScriptum GmbH & Co. KG
Alle Rechte vorbehalten. / All rights reserved. Saarbrücken 2008

Contents

1. **Location Models**   3
   1.1. Continuous Models 3
       1.1.1. Fermat-Weber Problem 3
       1.1.2. Geometric Steiner Tree Problem 5
       1.1.3. Covering Problems on Spatial Networks 5
   1.2. Discrete Models 6
       1.2.1. Minisum Problems 6
       1.2.2. Minimax Problems 6
       1.2.3. Double Coverage Aspects 8
       1.2.4. Art Gallery Problem 9
       1.2.5. Combination with Routing Problems 11

2. **Stochasticity in a Dynamic Environment**   13
   2.1. Stochastic Dynamic Warehouse Location Problem 13
   2.2. Exact Solution Method 16
       2.2.1. Stochastic Dynamic Programming 16
       2.2.2. Application to the SDFLP 17
   2.3. Heuristic Approach 18
       2.3.1. Results 21
   2.4. Conclusion and Further Research 21

3. **The Double Set Cover Problem**   25
   3.1. Introduction 26
   3.2. Optimization Models for the DSCP 28
   3.3. Complexity Results 31
       3.3.1. Complexity Results for the DSCP 31
       3.3.2. Complexity Results for the Optimization Versions of the DSCP 49
   3.4. Inference of Gene Regulatory Networks 59
       3.4.1. Introduction 59

Contents

- 3.4.2. The Gene Regulatory Network Problem 62
- 3.4.3. A Mixed-Integer Linear Programming Formulation 64
- 3.4.4. Complexity of the GRNP 68
- 3.4.5. Using ACO for solving the GRNP 70
- 3.4.6. Computational Experience 74
- 3.4.7. Randomly Generated Instances 74
- 3.4.8. A Real World Problem 77
- 3.4.9. Conclusion 81
- 3.5. Art Gallery Problems 82
  - 3.5.1. Complexity Issues 83
  - 3.5.2. Bounds 89
  - 3.5.3. Solution Techniques 93

A. Notions from Graph Theory and Computational Geometry 97
- A.1. Graph Theory 97
- A.2. Computational Geometry 99

B. NP-complete and NP-hard 101
- B.1. Complexity Theory 101
- B.2. Selected Problems 102

C. Experiments and Randomly Generated Instances 105
- C.1. Randomly Generated DCP Instances 105
- C.2. Experimenting on ODSCP1 and ODSCP2 108
- C.3. SDWLP: Instances and Solutions 110

1. Location Models

The field of location problems is wide and a lot has been done on this subject. In general, the problem is to locate facilities by choosing from a set of possible positions while respecting the effort (e.g.: costs) and the utility (e.g.: maintainance of a service). If the set is finite or countable we speak of a discrete problem and if we select the facility from a continuous set then the problem is continuous. For the discrete case the elementary problem is the *Set Cover Problem* (SCP), in which a finite family of finite sets is given. The problem is to find a subset of this family with a certain size that covers all elements. This problem is known to be NP-hard. The uncapacitated warehouse location problem (WLP) is a generalization of the SCP and a fundamental location problem. Here we have a number of warehouse locations and customers. We have to build warehouses (fixed cost) that supply the customers with a certain cost which is proportional to the the amount and the transportation distance. Many algorithms for solving this problem take advantage of the fact that fixing the locations leads to a simple transportation problem - which can be solved very efficiently. In this work we summarize mainly discrete models covered by the location theory to give an overview of the related literature. The central part is given by 2 new extensions: One concerning stochasticity in a dynamic environment [101] and another concerning a double Set Cover Problem plus applications, e.g. [82].

Most of the problems presented in the next two sections can be found in books that introduce to facility location. E.g.: [31] [27] and [41]. We will discuss minisum and minimax problems aswell as discrete and continuous ones. We put our focus on problems that are related to the extensions proposed in section 3 and section 2. We start with the presentation of some continuous models:

1.1. Continuous Models

1.1.1. Fermat-Weber Problem

One of the earliest studied location models is the Weber Problem, where a set of customers has to be served by a set of facilities(warehouses) while minimizing the sum of the weighted

1. Location Models

distances between customers and locations (e.g. d_{ij} distance). For the unweighted case with 3 customers the problem can be solved geometrically by constructing the so called Torricelli Point. For a nice geometric proof see [17]. The *Varignon Frame* is a mechanical device to find a solution to the Weber Problem. Figure 1.1 shows the functionality.

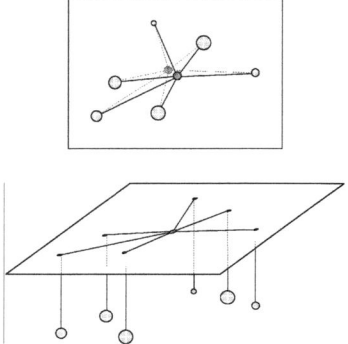

Figure 1.1.: Varignon frame

Weiszfeld [125] proposes a fix point iteration method to solve the problem with linear convergence. Li proposes a Newton acceleration [79] for the Weiszfeld Algorithm. The concept is also applicable for the multi-facility location problem. I.e.: We have n customers (position $c_i \in \mathbb{R}^2$) and we want to locate m facilities ($x_j \in \mathbb{R}^2$). Our aim is minimize the weighted distances between facilities and customers. I.e:

$$\sum_{i=1}^{n}\sum_{j=1}^{m} w_{ij}||c_i - x_j|| + \sum_{j=1}^{m}\sum_{k=j+1|}^{m} v_{jk}||x_j - x_k||$$

The Miehle's [87] algorithm solves this kind of problem and can be seen as a generalization of the Weiszfeld algorithm. The paper of Rosen [102] proposes an acceleration of Miehle's algorithm.

Changing the distance measure from euclidean to squared euclidean makes the problem easy since we can solve the problem analytically. For the one facility case the centroid (physical interpretation: center of gravity) solves the problem:

$$\mathbf{S} = \frac{1}{n}\sum_{i=1}^{n} w_i c_i$$

1.1.2. Geometric Steiner Tree Problem

A generalization of this problem is the *euclidean* and the *rectilinear* Steiner tree problem (planar Steiner tree problem). The geometric Steiner tree problem asks to connect points $P \subset \mathbb{R}^2$ in the plane with a tree of length L or less using terminals $Q \subset \mathbb{R}^2$. In the case of the *euclidean* STP we use the euclidean distance to measure the length of the tree and in the case of the *rectilinear* STP the Manhattan distance is used. An interesting counterpart to the rectilinear STP is the *Minimum Manhattan Network Problem*. Instead of minimizing the total length (with obviously results in a tree) we search for a Network that realizes the minimum distance between all pairs, regarding the Manhattan distance. For a approximations see [55] [6] and for an exact method based on a multi commodity flow formulation see [5]. Another variant generalizes the problem to the space [84] which has applications in nanotechnology. NP-completeness results for Steiner tree problems can be found in [45] and [44]. For a discussion of exact algorithms see [124]. The *euclidean STP* can be solved heuristically by using Delaunay Triangualtions.

1.1.3. Covering Problems on Spatial Networks

Covering Problems are simplified minimax problems. E.g.: we have to put a minimum number of centers in a way that all customers i are reachable in a maximum time s_i. We allow to locate the points X anywhere in the network - e.g.: we can also put a center in the middle of an edge. Since we define $d(X, i)$ as the minimum distance between the set X and the point i, the problem may be written in the following way:

$$\min |X|$$
$$d(X, i) \leq s_i.$$

Networks that have a tree structure are easy to solve [41]: By starting a search for centers we select a leaf node i. Then we check if the the adjacent node j is in the reach s_i. If this point is out of reach then we put a center on the edge - as far from node i as possible. Then we reduce the graph and eliminate all points that are reachable from this center. The graph remains a tree and we can restart the procedure. If we can reach j within s_i then we update $s_j \leftarrow \min\{s_j, s_i - d(i,j)\}$ and eliminate i. For cyclic networks the problem is hard to solve since it is equivalent to the discrete *Set Cover Problem* problem.

1.2. Discrete Models

1.2.1. Minisum Problems

Minisum problems are motivated by the question on locating a facility in a way that it minimizes the cost to service a set of customers. For a presentation of these problems see [13], [27], [41] and [51]. A standard problem is the placement of facilities in a way that the sum of transportation costs to all customers is minimized (transport). The most simple variant is finding the *median* $v^* \in V$ of a weighted undirected graph $G(V, E, c, b)$ which minimizes the sum of weighted distances to all other vertices. The weights can be interpreted as demands $b : V \to \mathbb{R}_0^+$ and $c : E \to \mathbb{R}_0^+$ can be seen as the distance of the corresponding arc. We define $d : V \times V \to \mathbb{R}_0^+$ as the length of the shortest path between two vertices. Then the median v^* is defined by:

$$\sum_{v \neq v^*} b_v d_{v^* v} = \min_{w \in V} \sum_{v \neq w} b_v d_{wv}$$

Hakimi[56] shows that it is not necessary to consider points located on the edges, which is also true for the *p-median* problem. The *p-median* $V^* \subset V$ is a subset of size p that minimizes the sum of minimum possible weighted distances. I.e:

$$\sum_{v \in V} b_v \min_{w \in V^*} \{d_{w,v}\} = \min_{W \subset V, |W|=p} \sum_{v \in V} b_v \min_{w \in W} \{d_{w,v}\}$$

Therefore the *p-median* provides a partition of $V = \bigcup_{i=1}^{p} V_i$ into p subsets. If the graph has the structure of a forest, then the problem gets easy to solve by successive decomposition into subtrees [52]. For the *Warehouse Location Problem* we choose from potential locations $i \in \{1, \ldots n\}$ ($y_i \in \{0, 1\}$) and give a transportation corresponding plan $x_{ij} \geq 0$ that satisfies the demand b_j for all customers $j \in \{1 \ldots m\}$. The costs associated to this decisions have to be minimized. Therefore we ask for a plan that minimizes the sum of the transportation costs $\sum \sum x_{ij} c_{ij}$ and fixed costs $\sum f_i y_i$. This kind of problem can be tackled with dual-based procedures originating in the method proposed in [33] as well as Branch & Bound and Lagrangian Relaxation. For large instances Genetic Algorithms are efficient heuristic solution methods, see [7] and [19].

1.2.2. Minimax Problems

Given a undirected weighted graph $G(V, E, c, b)$ with vertex weights $b : V \to \mathbb{R}_0^+$ and edge weights $c : E \to \mathbb{R}_0^+$. Let X be a set of p nodes, then our aim is to minimize the maximum distance beween a node $i \in V$ and a set of nodes X, i.e:

$$\min_{|X|=p} \max_{i \in V} d(X,i)$$

Again, if the graph is a tree the problem is easy to solve, but in general it is NP-hard. A simplification of the p-center problem is the Set Cover Problem SCP: In graph theoretical terms we have a given graph G(V,E) and we need to select a subset $X \subset V$ of p nodes in such a way that each node $i \in V$ is reachable within a maximum allowed distance from at least one node of X. Or, to formulate it with sets: given a family of sets $\mathcal{F}$ that covers a set S (i.e.: $\bigcup_{F \in \mathcal{F}} F = S$) we have to select a subset $\mathcal{G} \subset \mathcal{F}$ of limited size $|\mathcal{G}| = p$ that also covers S (i.e.: $\bigcup_{G \in \mathcal{G}} G = S$). In the weighted case each subset $F \in \mathcal{F}$ has a corresponding weight w_F and the restriction to pick p sets is replaced by a capacity constraint:

$$\sum_{F \in \mathcal{F}} w_F \leq C$$

Already this problem is NP-complete and there are is a lot of research related to exact and approxiamte solution procedures. Feige [35] shows a dramatic SCP inapproximability result, although SCP is "quite" easy in pratice. Therefore SCP subproblems are sometimes used as subproblem in heuristic approaches for p-center problem. An optimization version of the problem is the Maximum Covering Problem [107], for a heurtistic solution approach see [98]. If $\mathcal{F}$ consists of $F_i \in \mathcal{F}$ and the matrix $A=(a_{ij})$ is defined by $a_{ij} = 1 \Leftrightarrow j \in F_i$ ($i \in W$, $j \in V$) then MCP can be stated as the following IP:

$$\max \sum_{j \in V} w_j y_j$$
$$\text{s.t.} \sum_{i \in W} a_{ij} z_i \geq y_j$$
$$\sum_{i \in W} z_i = p$$
$$z_i \in \{0,1\}$$
$$y_j \in [0,1]$$

Center problems are often applied for problems in the public sector - for locating services like ambulances or fire departments. In this examples covering should also take care about the quality of the covering. Since the classical center problems suppose that the problem is static and that the capacity of the centers is unlimited. It is clear that these models need extensions

1. Location Models

to take these issues into account. In the next section we present how problems like congestions get handled.

1.2.3. Double Coverage Aspects

Double coverage aspects ocurr naturally in location problems in various forms. The main intend is to avoid drop outs of a service. If we think of a sercive that deals with emergencies then the demand is usually nondeterministic. Locating fire department is such a case: if we also suppose that the incidence of a demand (fire) is seldom, then two simultaneous incidences are very unlikely. Therefore we are instigated to think that it is practically impossible to have three incidences at the same time. Therefore we only consider tree cases - no incidence, one incidence and two incidences. In the case of *Ambulance Location Problems* we have to locate a fleet of p vehicles on m locations to cover all customers in a radius R while we also want maximize the proportion of those customers that are covered in smaller radius $r \leq R$ twice. The paer [49] deals with a Tabu Search for this problem. A variant that intends to discover more balanced solutions is described in [26]. A subproblem of the *Ambulance Location Problems* is the *Backup Coverage* problem decribed in [60]. Here our aim is to cover a given set of nodes in a way that all of them are covered at least once and as many as possible twice. The weighted variant can be stated as follows:

$$\max \sum_{j \in V} w_j y_j$$
$$\text{s.t.} \sum_{i \in W} a_{ij} z_i \geq 1 + y_j$$
$$\sum_{i \in W} z_i = p$$
$$z_i \geq y_i$$
$$y_j \in [0, 1]$$
$$z_i \in \{0, 1\}$$

Variants of *backup covering* problems can be also applied to security monitoring see [89], which also leads us to section 1.2.4.

1.2.4. Art Gallery Problem

The task of *Art Gallery Problems* is to position cameras to monitor a certain area. This problem is due to Victor Klee (in [63]) and has many different variations. We discuss the *Vertex Guard Problem* and the *Point Guard Problem*. For other variants we refer to [92] [112] and [121]. In the case of *Vertex Guard Problems* VGP we have to versions depending on the area that has to be monitored:

- select vertices (corner points or vertex guards) from a polygonal region (gallery) in such a way that each point on the boundary of the polygon is visible from at least one of the selected vertices.

- select vertices in such a way that each point of the polygon is visible from at least one of the selected vertices.

If we allow to position the guards anywhere we speak of a *Point Guard Problem*. Again, dependent on the area that has to be monitored we get two version. If a selection of vertices solves the VGP then we also speak of a *Vertex Guard Covering*.

For these problems we get a sharp bound by applying polygon triangulation [40]. I.e.: every triangulation is a 3-coloring of the vertices and therefore each color gives a feasible solution. Since it is impossible to have more than $\frac{n}{3}$ nodes in each color at least one color has less or equal $\frac{n}{3}$ nodes. That shows the that at least $\lfloor \frac{n}{3} \rfloor$ vertices are needed and this result is known as *Chvátal's Art Gallery Theorem*.

We emphasize that covering every point of the polygon is not equivalent to covering the whole border. The example given in Figure 1.2 shows that. Here we need to use 3 corner points to cover the border, but as we can see there is still a triangle in the middle that we can't reach. therefore we conclude that at least four corner points are needed to cover the whole area. By changing the example a bit (make the alcoves a bit more narrow) we can also see that this observation remains true if we allow to position the camera at any point of the polygon.

The proof presented above is due to [40] and makes use of triangulations. By giving an example (compare Figure 1.3) we show that the choice of the triangle may change the quality of the bound. I.e: in the first triangulation the number of nodes that have the same color are larger than 3 for all the colors and therefore we get a bound of 3. In the second case we obtain a better solution, which is in deed also optimal. We can see that the quality of the upper bounds are dependent on the triangulation.

The next example from Figure 1.4 shows that putting cameras only in the corner points is restrictive and we can see that allowing cameras somewhere on the border leads to improvements.

1. Location Models

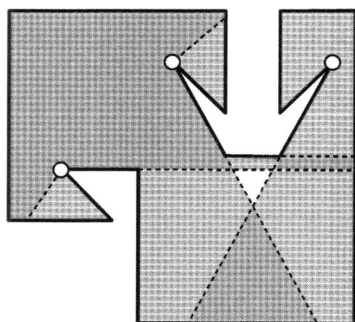

Figure 1.2.: Example where 3 cameras are sufficient to cover the border, but 4 cameras are needed to cover the whole area

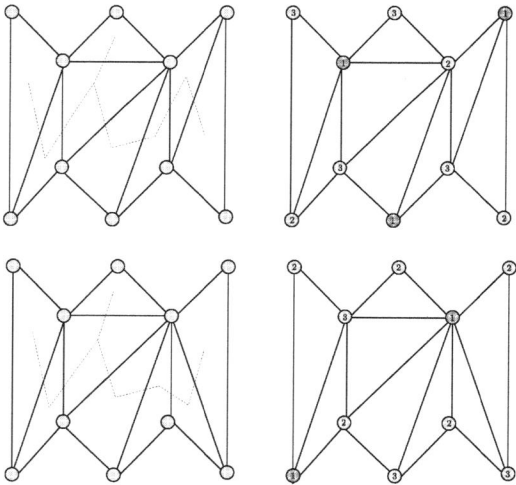

Figure 1.3.: Example of 2 different triangulations

1.2. Discrete Models

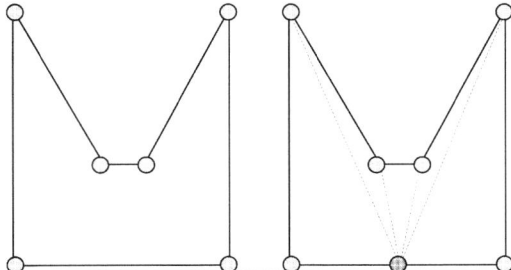

Figure 1.4.: An example where putting the camera on the wall leads to a better solution

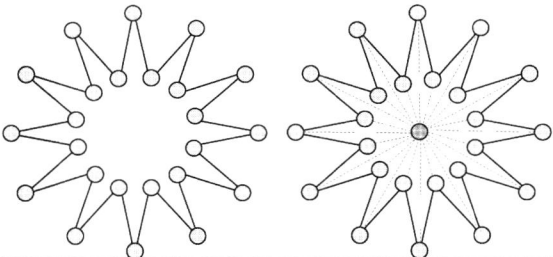

Figure 1.5.: An example where putting the camera inside the polygon leads to a better solution

The next step is allowing the camera somewhere in the room, and again it is possible to give an example where this leads to a better solution. In the example given in Figure 1.5 we can see that putting the camera in the middle of the star gives the optimal solution.

The *Minimum Vertex Guard Problem* is NP-hard (see [81]) for a approximation scheme see [53]. Inapproximability results can be found in [32]. After some preprocessing it is possible to use a SCP formulation to solve the problem exactly. The rectilinear variant of the problem is also NP-hard [109] and for exact method see [20].

1.2.5. Combination with Routing Problems

Suppost that we have to locate depots that serve as centers for a vehicle fleet. The basic variant is an extension of multi depot vehicle routing problem that connects the problem of locating a facility with routing decisions. I.e.: Location decisions affect the transportation costs and therefore it has to be taken into accout. For a review of location-routing problems see [90] and [88]. Especially for larger instances heuristics make use of clustering thechniques to get

11

1. Location Models

reasonable starting solutions. There are various types of location-routing problems and - just to mention a few - there exist dynamic versions [73], stochastic variants [75] and formulations that cover planar aspects [104].

2. Stochasticity in a Dynamic Environment

In this section we present a location model that respects dynamic and stochstic influences - it is a joint work with R. F. Hartl [101]. The Uncapacitated Facility Location Problem (UFLP) has been enhanced into many directions. In [91], [116] and [34] you can find numerous approaches that consider either dynamic or stochastic aspects of location problems. An exact solution method to an UFLP with stochastic demands is discussed in [74]. The problem considered there could be interpreted as a two stage stochastic programm. In [86] you can find dynamic (multi period) aspects as well as the multi commodity aspect. The approach in [105] could be seen as the integration of stochastics into the UFLP. In the work in hand a model will be presented, where the UFLP gets enriched by inventory and randomness in the demand. The UFLP and its generalizations are part of the class of NP-hard problems, where no exact efficient solution methods are known. First of all, the aim of this work is the preparation of tools to develop and investigate heuristics for this problem type. For this reason, an exact method for small instances was developed. This makes possible both, to carve out the range of exact solvability and to compare exact and heuristic solutions. A more detailed description of the problem is now following.

2.1. Stochastic Dynamic Warehouse Location Problem

Our aim is to find the optimal decisions for production, inventory and transportation, to serve the customers during a certain number of periods, $t \in \{1, ..., T\}$. Assume that the company runs a number for the production sites $i \in I = \{1, 2, ..., n\}$ that have limited storage capacities, $\Delta_i^{(t)}$. These production sites need not be used in all periods. When a production site i is operated at time t, this is denoted by the binary variable $\delta_i^{(t)} = 1$. In this case the fixed costs $o_i^{(t)}$ arise. If a location is active, then the exact production quantity $u_i^{(t)}$ must be fixed. For each period, the production decision is the first stage of the decision process. It has to be done before the demand of the customers is known. Only the current level of inventory $y_i^{(t-1)}$ as well as the demand forecasts are known in advance.

2. Stochasticity in a Dynamic Environment

Demand occurs at various customer locations $j \in J = \{1, 2, ..., m\}$. At any given period t the demand d_j^t at customer j will occur with probability p_j^t, whereas customer j will not require any delivery with probability $1 - p_j^t$. Hence, demand can be described by a dichotomous random variable[1] $\mathcal{D}_j^{(\tau)}$ ($\tau \geq t$). We also assume that the random variables $\mathcal{D}_j^{(t)}$ are stochastically independent.

$$\mathbf{P}(\mathcal{D}_j^{(t)} = d_j^{(t)}) = p_j^{(t)} \quad \mathbf{P}(\mathcal{D}_j^{(t)} = 0) = 1 - p_j^{(t)}$$

In the second stage, when the demand is known, we must decide upon the transportation of appropriate quantities $x_{ij}(t)$ from the production sites i to the customers j. We assume that the time needed for transportation can be neglected (i.e. the transportation lead time is less than one period). Stockouts (shortages) $f_j^{(t)}$ are permitted and are penalized by shortage costs $p_j^{(t)}$ per unit time and per unit of the product. We assume that backordering is not possible and that these potential sales are lost.

The periods are linked by the inventories $y_j^{(t)}$ at the production sites and the usual inventory balance equations (2.1) apply. Here $\eta_i^{(t)}$ denotes the surplus in period t at site i.

$$y_i^{(t)} + \eta_i^{(t)} = y_i^{(t-1)} + u_i^{(t)} - \sum_{j \in J} x_{ij}^{(t)} \tag{2.1}$$

In this section we assume free disposal, therefore the variable $\eta_i^{(t)}$ can be eliminated by turning the equality (2.1) into the inequality (2.2).

$$y_i^{(t)} \leq y_i^{(t-1)} + u_i^{(t)} - \sum_{j \in J} x_{ij}^{(t)} \tag{2.2}$$

After the completion of the production and transportation decisions and after updating the inventories, the next period can be considered. We note here, that for all periods we have to pay attention to the capacity restrictions (2.3).

$$0 \leq u_i^t + y_i^{(t-1)} \leq \Delta_i^{(t)} \tag{2.3}$$

In order to have a convenient notation, we introduce the concept of scenarios. A scenario $D_t \subset J$ is a subset of customers where the demand gets realized. Since the demands of the different customers are independent, the corresponding probability of a scenario to occur is given in formula (2.4).

---

[1] The embedding of stochastics is similar to the embedding of stochastics into the TSP, see [65].

2.1. Stochastic Dynamic Warehouse Location Problem

$$\mathbf{P}(D_t) = \prod_{j \in D_t} p_j^{(t)} \prod_{j \notin D_t} \left(1 - p_j^{(t)}\right) \qquad (2.4)$$

Solving the SDFLP means finding a strategy that minimizes the expected costs. Because of the sequencing of the decisions and the uncertain demand, the solutions could be understood as scenario dependent strategies, where the decisions are dependent on the forecasts and the level of inventory at hand. Figures 2.1(a) and 2.1(b) illustrate the dependency of operative planning[2] and the scenarios (realization of demand).

The left hand side of Figure 2.1(a) shows the production decisions $u_i^{(t)}$ and all of the possible subsequent scenarios (8 in number). One of the scenarios is magnified in the upper part of Figure 2.1(b). In each scenario the decisions for transportation, inventory and shortage are necessary.

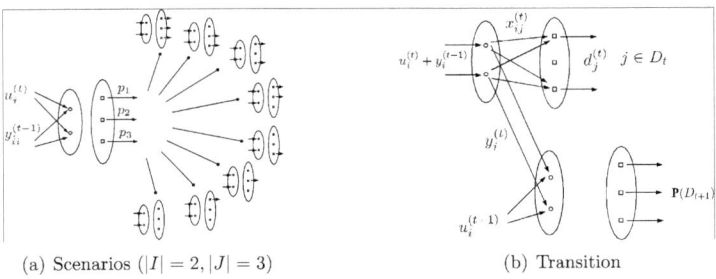

(a) Scenarios ($|I| = 2, |J| = 3$)   (b) Transition

Figure 2.1.: Sequencing of decisions

In order to complete the model formulation, we summarize the decision variables and the corresponding costs in Table 2.1.

The decisions $\delta_i^{(t)}$ and $u_i^{(t)}$ are linked by formula (2.5)

$$\delta_i^{(t)} = \begin{cases} 1 & \text{if } u_i^{(t)} > 0 \\ 0 & \text{if } u_i^{(t)} = 0 \end{cases} \qquad (2.5)$$

while shortages are defined as

$$f_j^{(t)} = \mathcal{D}_j^{(t)} - \sum_{i \in I} x_{ij}^{(t)}.$$

The total cost F is the sum over all periods of fixed operating costs, variable production costs.

---
[2]to keep the figure as simple as possible shortages and disposal are not integrated.

2. Stochasticity in a Dynamic Environment

Table 2.1.: Variables and costs

variable	cost	description
$\delta_i^{(t)} \in \{0,1\}$	$o_i^{(t)}$	operating decision and fixed costs
$x_{ij}^{(t)} \in \mathbf{Z}_+$	$c_{ij}^{(t)}$	transportation decision and unit transportation cost
$y_i^{(t)} \in \mathbf{Z}_+$	$s_i^{(t)}$	inventory level and unit holding cost
$u_i^{(t)} \in \mathbf{Z}_+$	$m_i^{(t)}$	production decision and variable production cost
$f_j^{(t)} \in \mathbf{Z}_+$	$p_j^{(t)}$	shortage (lost sales) and unit shortage cost

$$F = \mathbf{E}\left(\sum_{t=1}^{T}\sum_{i \in I}\left[o_i^{(t)}\delta_i^{(t)} + m_i^{(t)}u_i^{(t)} + s_i^{(t)}y_i^{(t)}\right] + \sum_{t=1}^{T}\sum_{i \in I}\sum_{j \in J} c_{ij}^{(t)}x_{ij}^{(t)} + \sum_{t=1}^{T}\sum_{j \in J} p_j^{(t)}f_j^{(t)}\right)$$

Since all relevant information about the past is contained in the inventory levels, this model is well suited to be solved by dynamic programming. This will be outlined in the next section.

2.2. Exact Solution Method

2.2.1. Stochastic Dynamic Programming

The principle of dynamic programming is the recursive estimation of the value function. This value function, henceforward denoted by F, contains the aggregate value of the optimal costs in all remaining periods. It can be derived recursively. It is convenient to first describe the method in general and to apply it to the problem afterwards. Let $z \in \mathbf{R}_+^m$ be the vector of state variables and $u \in \mathbf{R}_+^n$ be the vector of decisions. The set of feasible decisions in state z and period t is denoted by $U_t(z)$. The random influence in period t is represented by the random vector $r^{(t)}$ for which the corresponding distribution is known. It is important to note that the random variables $\{r^{(t)}\}$ have to be stochastically independent. The state transformation is described by

$$z_{t+1} = A(z_t, u_t, r_t)$$

and depends on the current state z_t, the random influence r_t at time t, and the chosen decision u_t. The single period costs in period t and state z when decision u is taken and random variable

2.2. Exact Solution Method

r is realized is denoted by $g_t(z, u, r)$.

The value function $F_t(z)$ gives the minimal expected remaining costs when starting in state z in period t. We now present a variant of the stochastic Bellman equation (compare Schneeweiß[108] S.151 (10.25) or Bertsekas [4] S.16).

$$\begin{aligned} F_T(z) &= \min_{u \in U_T(z)} \left\{ \mathbf{E}\left[g_T(z, u, r^{(T)})\right] \right\} \\ F_t(z) &= \min_{u \in U_t(z)} \left\{ \mathbf{E}\left[g_t(z, u, r^{(t)}) + F_{t+1}(A_t(z, u, r^{(t)}))\right] \right\} \quad t = T-1, \ldots, 1 \end{aligned} \quad (2.6)$$

Recursively solving the equation (2.6) we get an optimal strategy that balances the cost for implementing the decision u and the expected resulting remaining costs.

Applying Stochastic Dynamic Programming (SDP) to the SDFLP is almost straight forward. For solving the problem we have to iteratively calculate the functions F_t. We will show later that for the SDFLP it is sufficient to consider integer controls.

2.2.2. Application to the SDFLP

In order to apply the DP equation (2.6) to the SDFLP, we first introduce the notation $G(D, y_i^{start}, y_i^{end}, t)$ for the sum of inventory holding costs, shortage costs, and transportation costs in scenario D in period t when starting with initial inventory levels y_i^{start} and where the final inventories are required to be y_i^{end}. To every given inventory level y_i^{start} and scenario D, the best possible transportation plan has to be calculated. This can be done by solving a linear program:

$$G(D, y_i^{start}, y_i^{end}, t) = \sum_{i \in I} s_i^{(t)} y_i^{end} + \min_{x_{ij}, f_j} \sum_{j \in J} p_j^{(t)} f_j + \sum_{i \in I} \sum_{j \in J} c_{ij}^{(t)} x_{ij} \quad (2.7)$$

$$\begin{aligned} \text{s.t.} \quad & \sum_{i \in I} x_{ij} + f_j = d_j^{(t)} && \forall j \in D \\ & \sum_{j \in J} x_{ij} + y_i^{end} \le y_i^{start} && \forall i \in I \\ & f_j, x_{ij} \ge 0. \end{aligned}$$

Now the value function F_T of the final period T can be computed. In G, the starting inventory is now given by $y_i^{start} = u_i^{(T)} + y_i^{(T-1)}$ while the terminal inventory must be zero, $y_i^{end} = 0$:

2. Stochasticity in a Dynamic Environment

$$F_T(y_i^{(T-1)}) = \min_{\substack{u_i^{(T)} \geq 0 \\ 0 \leq u_i^T + y_i^{(T-1)} \leq \Delta_i^{(T)}}} \left\{ \sum_{i \in I} \delta_i^{(T)} o_i^{(T)} + \sum_{i \in I} u_i^{(T)} m_i^{(T)} + \sum_{D_T \subset J} \mathbf{P}(D_T) G(D_T, u_i^{(T)} + y_i^{(T-1)}, 0, T) \right\} \quad (2.8)$$

Going back in time, we have to turn to the general case in period $t < T$. Now we have to take into account the remaining costs in periods $t+1, ..., T$ when making the decision in period t. The starting inventory is now given by $y_i^{start} = u_i^{(t)} + y_i^{(t-1)}$ while the inventory at the end of period t is $y_i^{end} = y_i^{(t)}$. When determining $G(D_t, u_i^{(t)} + y_i^{(t-1)}, y_i^{(t)}, t)$ again a linear program has to be solved. The recursion for the value function becomes:

$$F_t(y_i^{(t-1)}) = \min_{\substack{u_i^{(t)} \geq 0 \\ 0 \leq u_i^{(t)} + y_i^{(t-1)} \leq \Delta_i^{(t)}}} \left\{ \sum_{i \in I} \delta_i^{(t)} o_i^{(t)} + u_i^{(t)} m_i^{(t)} + \sum_{D_t \subset J} \mathbf{P}(D_t) \min_{0 \leq y_i^{(t)} \leq \Delta_i^{(t+1)}} \left\{ G_t(D_t, u_i^{(t)} + y_i^{(t-1)}, y_i^{(t)}, t) + F_{t+1}(y_i^{(t)}) \right\} \right\} \quad (2.9)$$

In the SDFLP the data and the controls are assumed to be integer. In the problem (2.7) we therefore have to solve an integer linear program. It turns out to be a min cost flow problem and therefore it is totally unimodular, such that using the Simplex method for solving the relaxed linear program results in integer solutions for the transportation quantities x_{ij} and the shortages f_j.

The computational effort of this exact algorithm is increasing exponentially with the capacity at the locations and the number of customers. The additional effort that emerges from adding additional periods to the problem is linear.

This DP formulation is only applicable for small problem instances and for larger problem instances heuristic approaches are necessary. This is considered in the next section.

2.3. Heuristic Approach

A heuristic designed to solve stochastic combinatorial optimization problems is the Sample Average Approximation Method (SAA); see Kleywegt et al. [71]. Our model deals with a multi stage problem and that is the reason why this method is not directly applicable. In what follows we first present the classical SAA for solving static stochastic combinatorial optimization problems. Afterwards, we will explain how this method can be modified in order to be applicable

2.3. Heuristic Approach

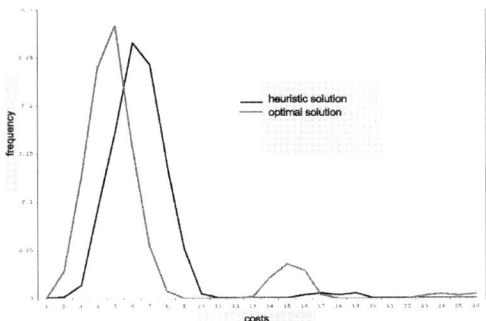

Figure 2.2.: Comparison of different solutions

to our problem.

Consider the following stochastic combinatorial optimization problem (2.10) in which W is a random vector with known distribution P, and S is the finite set of feasible solutions.

$$v^\star = \min_{x \in S} g(x), \quad g(x) := \mathbf{E}_P G(x, W) \tag{2.10}$$

The main idea of the SAA method is to replace the expected value $\mathbf{E}_P G(x, W) = \int G(x, w) P(dw)$ (which is usually very time consuming) by the average of a sample. The following substitute problem (2.11) is an estimator of the original problem (2.10).

$$\min_{x \in S} \hat{g}_N(x), \quad \hat{g}_N(x) := \frac{1}{N} \sum_{j=1}^{N} G(x, W^j) \tag{2.11}$$

The SAA method works in three steps

1. Generate a set of independent identically distributed samples: $\{W_i^1, \ldots, W_i^N\}_{i=1}^{M}$ of the random variable W.

2. Solve the corresponding optimization problems, i.e. optimize:

$$\hat{v}_i = \min_{x \in S} \hat{g}_i(x), \quad \hat{g}_i(x) = \frac{1}{N} \sum_{j=1}^{N} G(x, W_i^j).$$

3. Estimate the solution quality. This is done by first computing mean and variance of the

2. Stochasticity in a Dynamic Environment

sample:

$$\hat{v} = \frac{1}{M} \sum_{i=1}^{M} \hat{v}_i, \quad \hat{\sigma}^2 = \frac{1}{M(M-1)} \sum_{i=1}^{M} (\hat{v}_i - \hat{v})^2.$$

Then a solution $\tilde{x}$ is chosen (e.g. we can take the solution with the smallest $\hat{v}_i$) and its objective value is estimated more accurately by generating a larger sample $\{W^1, \ldots, W^{N'}\}$ ($N' \gg N$)

$$\tilde{v} = \frac{1}{N'} \sum_{j=1}^{N'} G(\tilde{x}, W^j), \quad \tilde{\sigma}^2 = \frac{1}{N'(N'-1)} \sum_{j=1}^{N'} (G(\tilde{x}, W^j) - \tilde{v})^2.$$

Calculate the value gap and σ_{gap}^2:

$$gap = \tilde{v} - \hat{v}, \quad \sigma_{gap}^2 = \tilde{\sigma}^2 + \hat{\sigma}^2$$

Since $\mathbf{E}(\hat{v}) \leq v^\star \leq \mathbf{E}(\tilde{v})$ the values $\tilde{v}$ and $\hat{v}$ can be interpreted as bounds on $v^\star$: let $x^\star \in S$ denote an optimal solutions of (2.10) then the first inequality $\mathbf{E}(\hat{v}) \leq v^\star$ comes from taking the expected value on the following inequality:

$$\hat{v}_i \leq \hat{g}_i(x^\star)$$

which results in

$$\mathbf{E}(\hat{v}_i) \leq \mathbf{E}(\hat{g}_i(x^\star)) = v^\star.$$

After completing Step 3, we have to inspect the values of gap and σ_{gap}^2. If these values are too large, one must repeat the procedure with increased values of N, M and N'. In [105] this method is applied for a Supply-Chain Management problem that includes location decisions.

Because of the multi-period structure of the SDFLP, the above SAA procedure has to be adapted. In particular, one must pay special attention to the way how the sampling is done. The sampling is done independently in every stage and every state of the SDP and we simply modifying formulas (2.8) and (2.9), where for every period and inventory level we only take the sum over a small randomly chosen sets of scenarios. To be more specific, the expected value in formula (2.9) passes over into (2.12) where $\{D_i\}$ denotes the sample chosen in stage t and state $y_i^{(t)}$.

$$\frac{1}{N} \sum_{i=1}^{N} \min_{0 \leq y_i^{(t)} \leq \Delta_i^{(t+1)}} \{ G_t(D_i, u_i^{(t)} + y_i^{(t-1)}, y_i^{(t)}, t) + F_{t+1}(y_i^{(t)}) \} \tag{2.12}$$

2.3.1. Results

The implementation was done in C++ and an additional library from the GNU Linear Programming Kit (GLPK) [83] was used. For small examples where the product $\prod_{i \in I} \Delta_i^{(t)}$ is small enough (say ≤ 50) it is possible to find the exact solution using the dynamic programming approach described in Section 3. In Figure 2.2, we can see the cost distributions for the exact and the heuristc approach in a small example (3 periods, 3 customers, 2 facilities and capacity = 3). For more information about the instance see section C.3. Here one can see that the shapes are quite different. The instance considered has very uncertain demand (every customer has probability 50%). Hence the two peaks in the optimal solution are not very surprising. It is interesting to observe that the heuristic solution does not show these twin peaks.

This second peak diminishes if the probability is close to 0 or 1. An experiment was made where the probability for the demand varied from 0 to 1 for all customers, i.e.: ($p_j^{(t)} = p \in \{0, 0.01, 0.02, \ldots, 1\}$). The result is depicted in Figure 2.3 where grey areas represent positive probabilities that these cost values occur. Every vertical line (p fixed) corresponds to a distribution function. For instance, at $p = 0.1$ five peaks occur. When the probability p increases, the number of peaks in the distribution function decreases.

The key decision to make the heuristic work well is to choose the right sample size N and the right number of samples M. In Figure 2.4 the statistical lower bound $\hat{v}$ (calculated in step 3) is depicted for different values of N and M. Choosing a sample size N that is large enough seems to be more important than a large number of samples. In Figure 2.4(b) the region $N > 70$ of Figure 2.4) is magnified to see the effect of a choosing the number of samples more clearly. One also can see that the statistical gap stays positive if at least 11 samples of size 71 are chosen. It is also interesting to note that for small values of N and sufficient large M the corresponding bounds are quite good, although the corresponding individual solutions are quite bad. This situation is depicted in Figure 2.5.

2.4. Conclusion and Further Research

In this section a stochastic dynamic facility location problem was proposed and exact and heuristic solution methods were presented. The examples that can be solved to optimality are quite small and therefore of minor practical interest. But the comparison of the SAA results and the exact solution method shows the applicability of the proposed method for larger instances of the SDFLP. To get more insight into this method, it will be necessary to make a transfer of the theoretical results known for the SAA method (see [71] for statistical bounds). For comparison purposes it would be interesting to adopt metaheuristic concepts: e.g. by using

2. Stochasticity in a Dynamic Environment

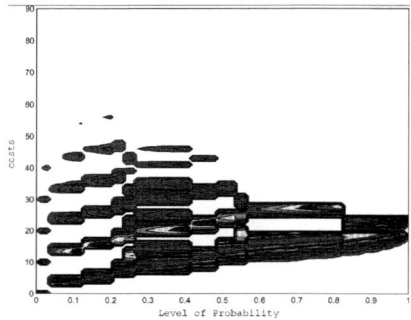

Figure 2.3.: Distributions of the optimal solution to instances with different levels of probability $(p_j^{(t)} = p \in \{0, 0.01, 0.02, \ldots, 1\})$

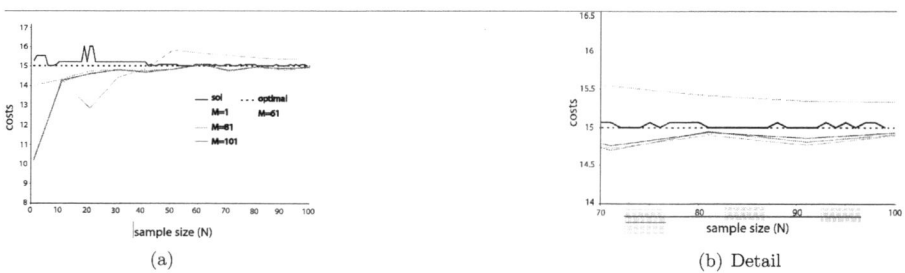

Figure 2.4.: Choice of sample size N and the number of samples M.

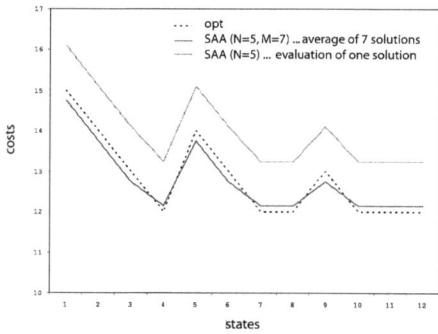

Figure 2.5.: Expected costs to different levels of inventory y (1 : $[0,0]$; 2 : $[0,1]$; ...; 4 : $[0,3]$; 5 : $[1,0]$; ...; 12 : $[2,3]$).

the variable-sample approach (for references see [62]). In our further research we also want to consider other exact solution techniques considering the SDFLP as a multistage stochastic program.

3. The Double Set Cover Problem

In this chapter we formulate the *Double Set Cover Problem* (DSCP). The DSCP is an extension of the SCP. Instead of one service, the DSCP intends to offer two different services that "compete" for possible centers. We will present applications of the DSCP and we investigate the complexity, i.e.: we search for the limit where DSCP changes from a NP-complete problem to an polynomially solvable one. Two different optimization versions of the DSCP will be given and a heuristic is developed to solve large instances. We give a real world application to resolve gene regulatory networks and we continue the discussing with a double cover variant of the *Vertex Guard Problem* for *Art Galleries*.

Before we start to formalize the DSCP we state the following problem: suppose that we have to maintain 2 services (e.g.: police departments, fire departments) in a sparsely populated region. And we have to locate facilities to implement that. There is a couple of places that may serve as a center, but dependent on the site it will only cover a certain region. Additionally we assume that these regions are not only dependent on the site, but also on the service. Furthermore, we want to choose at most k locations and we want to avoid to locate two facilities in one and the same location.

The examined DSCP is a generalization of the SCP and refers to the following problem:

- Given: a set S, a family $\mathcal{F} = \{(S_i^{(1)}, S_i^{(2)}) | S_i^{(1)}, S_i^{(2)} \subset S \wedge i = 1 \ldots m\}$ and an integer k.

- Find selections $\Lambda_1, \Lambda_2 \subset \{1 \ldots m\}$ such that:

$$|\Lambda_1 \cup \Lambda_2| \leq k \tag{3.1}$$

$$\Lambda_1 \cap \Lambda_2 = \phi \tag{3.2}$$

$$\bigcup_{i \in \Lambda_1} S_i^{(1)} = S \tag{3.3}$$

$$\bigcup_{i \in \Lambda_2} S_i^{(2)} = S \tag{3.4}$$

3. The Double Set Cover Problem

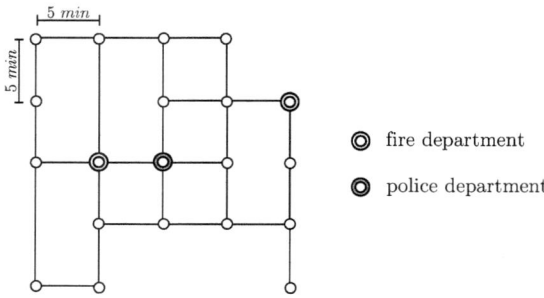

Figure 3.1.: Example for a road network

3.1. Introduction

To motivate the problem we give a short example. Again, we suppose that we have a couple of villages in a given area and we have to locate police and a fire departments to serve all the villages. Furthermore we suppose that each village has to be reached from a police department within a given maximum time and similarly each village also has to be reached by the firefighters within another given time limit. For instance 20 minutes for the police and 15 minutes for the fire department. In Figure 3.1 we can see a rectangular road network showing with a feasible solution, where only two fire departments and one police department are used.

In this context the problem extends the classical center problem since it concerns two types of services that "compete" for the locations. The exclusive usage of locations may have different reasons:

- *safety and convenience*: it may also be unconvenient or even dangerous to combine facilities in the same place. E.g: land fill, recreation area, airport etc.

- *security and utility*: overlapping duties or services. police, firefighters, ambulance, shops. positive effect: each service covers the whole region and we additionally get a double coverage of the intersecting services.

- *fairness*: villages may compete for the facilities.

- *space*: there may simply be insufficient space to build both facilities at the same place

For many real world problems it is necessary to add a budget constraint which intuitively generalizes the DSCP to the *Weighted Double Set Cover Problem* WDSCP. Here, every $i \in$

3.1. Introduction

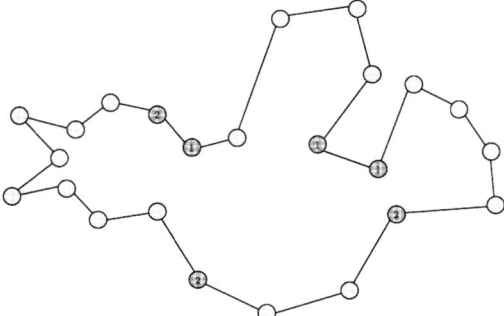

Figure 3.2.: Polygon with two independent surveillance systems

$\{1\ldots m\}$ has two different weights w_i^1 and w_i^2 that correspond to the service. I.e.: if we select $i \in \Lambda_1$ then it costs w_i^1 and if we select $i \in \Lambda_2$ then it costs w_i^2. Therefore we only replace (3.1) by

$$\sum_{i \in \Lambda_1} w_i^1 + \sum_{i \in \Lambda_2} w_i^2 \leq k$$

and we get the formulation for WDSCP. WE note that the MIP formualtions and many of the results will be presented that are also applicable to the WDSCP.

We continue with discussing the unweighted version and give another interpretation of the DSCP that is motiveated by *Art Gallery Problems*(see setion 1.2.4). More precisely, it is an extension of the *Vertex Guard Problem* for art galleries: Suppose that we want to install two independent surveillance systems for an art gallery. I.e.: the art gallery is again represented as a polygon and the cameras are installed in the corners of the room. The aim is to observe the whole border of the polygon. Both systems have to cover all edges and in one corner there is only space for one camera. We will show that this variant is NP-complete and we will give an upper bound for the number of cameras.

As a third application we consider a special kind of stratification problem. I.e: we deal with a two-shift production with n process types and m supervisors that control operations or in case react to occurring problems. Every process needs special knowledge to respond properly, therefore S are the tasks and for each supervisor i a set of permissible tasks $S_i = S_i^{(1)} = S_i^{(2)}$ is given. Since we don't allow working in two consecutive shifts we are searching for a double covering of the operation types. If the operation types in shift one $S^{(1)}$ differ from shift two $S^{(2)}$, we may also transform the problem into a DSCP, by adapting S, $S_i^{(1)}$ and $S_i^{(2)}$. I.e.:

$$S = S^{(1)} \cup S^{(2)}$$

27

3. The Double Set Cover Problem

$$S_i^{(1)} \leftarrow S_i^{(1)} \cup (S^{(2)} \setminus S^{(1)}) \quad S_i^{(2)} \leftarrow S_i^{(2)} \cup (S^{(1)} \setminus S^{(2)})$$

Now we formulate two optimization versions of the DSCP:

3.2. Optimization Models for the DSCP

We consider the optimization versions by penalizing infeasibilities. I.e.: we discuss the following two:

- $\bigcup_{i \in \Lambda_1} S_i^{(1)} \neq S$
- $\bigcup_{i \in \Lambda_2} S_i^{(2)} \neq S$

while the following condition holds:

- $\bigcup_{i \in \Lambda_1} S_i^{(1)} \cup \bigcup_{i \in \Lambda_2} S_i^{(2)} = S$

The aim is to maximize the number of nodes that get serviced regularly. I.e.:

$$\max | \bigcup_{i \in \Lambda_1} S_i^{(1)} \cap \bigcup_{i \in \Lambda_2} S_i^{(2)} |$$

or equivalently, to minimize the number of nodes that get serviced by only one type of facility. I.e.:

$$\min | \bigcup_{i \in \Lambda_1} S_i^{(1)} \triangle \bigcup_{i \in \Lambda_2} S_i^{(2)} |$$

Now, we will derive a linear model ODSCP1 for this kind of problem that uses the following decision variables that indicate the solution Λ_1 and Λ_2:

$$z_i^{(1)} = \begin{cases} 1 & \text{if } i \in \Lambda_1, \\ 0 & \text{otherwise.} \end{cases}$$

$$z_i^{(2)} = \begin{cases} 1 & \text{if } i \in \Lambda_2, \\ 0 & \text{otherwise.} \end{cases}$$

the auxiliary variables $\delta^{(1)}$ indicate the set $\bigcup_{i \in \Lambda_1} S_i^{(1)}$ while $\delta^{(2)}$ indicates the set $\bigcup_{i \in \Lambda_2} S_i^{(2)}$.

$$\delta_j^{(1)} = \begin{cases} 1 & \text{if } j \in \bigcup_{i \in \Lambda_1} S_i^{(1)} \\ 0 & \text{otherwise.} \end{cases}$$

$$\delta_j^{(2)} = \begin{cases} 1 & \text{if } j \in \bigcup_{i \in \Lambda_2} S_i^{(2)} \\ 0 & \text{otherwise.} \end{cases}$$

3.2. Optimization Models for the DSCP

Additionally we represent $\mathcal{F}$ by the two matrices $A_1 = (a_{ij}^{(1)})$ and $A_2 = (a_{ij}^{(2)})$, where:

$$a_{ij}^{(1)} = \begin{cases} 1 & \text{if } j \in S_i^{(1)} \\ 0 & \text{otherwise.} \end{cases} \qquad (3.5)$$

$$a_{ij}^{(2)} = \begin{cases} 1 & \text{if } j \in S_i^{(2)} \\ 0 & \text{otherwise.} \end{cases} \qquad (3.6)$$

The model as a whole can be stated as a MIP in the following way:

$$\underbrace{\min_{z,\delta} \sum_{i \in S} (1 - \delta_i^{(1)}) + (1 - \delta_i^{(2)})}_{f_{ODSCP1} :=}$$

$$\begin{aligned}
\text{s.t.} \quad & \sum_{i \in S} a_{ij}^{(1)} z_i^{(1)} \geq \delta_j^{(1)} && \forall j \in S & (3.7) \\
& \sum_{i \in S} a_{ij}^{(2)} z_i^{(2)} \geq \delta_j^{(2)} && \forall j \in S & (3.8) \\
& z_i^{(1)} + z_i^{(2)} \leq 1 && \forall i \in S & (3.9) \\
& \sum_{i \in S} z_i^{(1)} + z_i^{(2)} \leq k && & (3.10) \\
& \sum_{i \in S} a_{ij}^{(1)}(z_i^{(1)} + z_i^{(2)}) \geq 1 && \forall j \in S & (3.11) \\
& \sum_{i \in S} a_{ij}^{(2)}(z_i^{(1)} + z_i^{(2)}) \geq 1 && \forall j \in S & (3.12) \\
& z_i^{(1)}, z_i^{(2)} \in \{0,1\} && \forall i \in S & (3.13) \\
& \delta_i^{(1)}, \delta_i^{(2)} \in \{0,1\} && \forall i \in S & (3.14)
\end{aligned}$$

It is easy to see that: $\sum_{i \in S}(1 - \delta_i^{(1)}) = |S \setminus \bigcup_{i \in \Lambda_1} S_i^{(1)}|$ and $\sum_{i \in S}(1 - \delta_i^{(2)}) = |S \setminus \bigcup_{i \in \Lambda_2} S_i^{(2)}|$ and therefore $f_{ODSCP1} = |\bigcup_{i \in \Lambda_1} S_i^{(1)} \triangle \bigcup_{i \in \Lambda_2} S_i^{(2)}|$. We also note that the variables $\delta_i^{(1)}$ and $\delta_i^{(2)}$ may be relaxed, without loosing integrality in the optimal solution:

$$\delta_i^{(1)}, \delta_i^{(2)} \in [0,1] \quad \forall i \in S \qquad (3.15)$$

Another intuitive way to derive an optimization version is to penalize infeasibilities where $\Lambda_1 \cap \Lambda_2 \neq \phi$, i.e.: we minimize:

$$\min |\Lambda_1 \cap \Lambda_2|.$$

To formulate the corresponding MIP ODSCP2 we tie in with the formulation of ODSCP1. The difference of the formulations is the meaning of δ:

3. The Double Set Cover Problem

$$\delta_i = \begin{cases} 1 & \text{if } i \in \Lambda_1 \cap \Lambda_2 \\ 0 & \text{otherwise.} \end{cases}$$

Summing up, the formulation gets the following simple form:

$$\min_{z,\delta} \sum_{i \in S} \delta_i$$
$$f_{ODSCP2} :=$$

s.t.	$\sum_{i \in S} a_{ij}^{(1)} z_i^{(1)} \geq 1$	$\forall j \in S$	(3.16)
	$\sum_{i \in S} a_{ij}^{(2)} z_i^{(2)} \geq 1$	$\forall j \in S$	(3.17)
	$z_i^{(1)} + z_i^{(2)} \leq 1 + \delta_i$	$\forall i \in S$	(3.18)
	$\sum_{i \in S} z_i^{(1)} + z_i^{(2)} \leq k$		(3.19)
	$z_i^{(1)}, z_i^{(2)} \in \{0,1\}$	$\forall i \in S$	(3.20)
	$\delta_i \in \{0,1\}$	$\forall i \in S$	(3.21)

This formulation (3.16) may be regarded as a node-formulation while the second one is more like an edge-formulation (3.7). Computational experiments (compare section C.2) give reason to believe that the edge version leads to LP models that are easier to solve by LP solvers like CPLEX. One reason may be that finding solutions is an easier task. Let X_{ODSCP1} denote the solution space of ODSCP1 and X_{ODSCP2} the solution space of ODSCP2 then the following Lemma holds:

Lemma 3.2.1. $X_{ODSCP2} \subsetneq X_{ODSCP1}$.

Proof. Let $z = (z^{(1)}, z^{(2)}) \in X_{ODSCP2}$ be a feasible solution for ODSCP2 and suppose that $z_i^{(1)} + z_i^{(2)} = 2$, then i is a facility that holds both types of services. To make i feasible for ODSCP1 we set $z_i^{(2)} = 0$. Therefore we have to take care of the nodes j that were exclusively serviced by facility i and service type 2. To generate a feasible solution for ODSCP1 we set $\delta_j = 0$ for exactly those nodes, while the others are set to one. To show that generally $X_{ODSCP2} \neq X_{ODSCP1}$, it is sufficient to give an example. Therefore we set $k = 1$ and define $A_1 = A_2 = A$ and:

$$A := \begin{pmatrix} 1 & 1 \\ 0 & 1 \end{pmatrix}$$

For ODSCP1 we get a feasible solution if we set $z_1^{(1)} = 1$ and $\delta_1^{(2)} = \delta_2^{(2)} = 0$. In the case of ODSCP2 we are only allowed to label at most one node with exactly one of the services. So we can only provide one of the services and therefore it is impossible to construct a feasible solution. □

3.3. Complexity Results

In this section we present complexity results for DSCP and the optimization versions ODSCP1 and ODSCP2. These results incorporate information about the underlying structure of $\mathcal{F}$ and the density of A_1, A_2 respectively. First, we will simply show that the problem is NP complete and then we will continue with more detailed complexity results.

3.3.1. Complexity Results for the DSCP

Since the SCP is NP-complete and because of the relatedness of DSCP and SCP it seems obvious that also the DSCP is NP-complete - which is certainly true, but it is interesting to notice that the "double" aspect makes the problem harder to solve. In fact, we will show that the DSCP may still be hard to solve although the corresponding uncoupled SCPs for A_1 and A_2 are easy to solve. First, we will define the SCP and then we will proof that DSCP is NP-complete.

Definition Set Cover Problem (SCP). Given a finite set $S = \{1\ldots n\}$ and a collection of subsets of S, namely $\mathcal{F} = \{S_i \subset S | i = 1\ldots m \}$. Does $\mathcal{F}$ contain a cover of S of size k? Or equivalently, does a subset $\Lambda \subset \{1\ldots m\}$ of size k exist such that $S = \bigcup_{i \in \Lambda} S_i$?

Theorem 3.3.1. *DSCP is NP-complete.*

Proof. Given an instance of SCP $\mathcal{F} = \{S_i \subset S | i = 1\ldots m\}$ with a certain k ($k \leq |S|$), we will construct an instance $\tilde{\mathcal{F}} = \{(\tilde{S}_i^{(1)}, \tilde{S}_i^{(2)}) | i = 1\ldots m\}$ of DSCP:

$$\tilde{S}_i^{(1)} = S_i \quad \tilde{S}_i^{(2)} = S \quad \tilde{k} = k+1$$

If $\tilde{\mathcal{F}}$ contains a feasible double covering (Λ_1, Λ_2), then $|\Lambda_2| \neq \phi$ and $|\Lambda_2| \leq k$. Therefore $\Lambda := \Lambda_1$ solves the SCP. On the other hand, if Λ is a feasible set covering, then we define $\Lambda_1 := \Lambda$ and $\Lambda_2 := \{i'\}$ (any $i' \in S \setminus \Lambda$). Obviously (Λ_1, Λ_2) is a feasible double covering. □

Before we go on with more restrictive variants of the DSCP we note that any DSCP instance $\mathcal{F}$ can be characterized by the number k and the two zero-one matrices A_1 and A_2, compare (3.5) and (3.6). Furthermore, we point out that a zero-one matrix $A \in \{0,1\}^{m \times n}$ ($m \leq n$) may be interpreted as an adjacency matrix of a directed graph $G(V, E)$ since we may set: $V = \{1, \ldots, n\}$ and an arc (i,j) $\in E$ exists if and only if $a_{ij} = 1$.

3. The Double Set Cover Problem

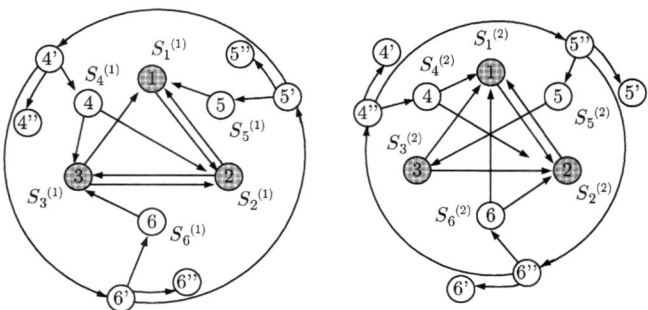

Figure 3.3.: Example of a transformation of $\mathcal{F} = \{(\{2\},\{2\}), (\{1,3\},\{1\}), (\{1,2\},\{1,2\}), (\{2,3\},\{1,2\}), (\{1\},\{3\}), (\{3\},\{1,2\})\}$

Definition Density. We let $A \in \{A_1, A_2\}$ and in accordance to the interpretation of a zero-one matrix A as a graph we define the row and the column density for DSCP instances.

$$d_A^+(i) := \sum_{j=1}^n a_{ij}$$

$$d_A^-(j) := \sum_{i=1}^m a_{ij}$$

Additionally we define:

$$d_A^+ := \max\{d_A^+(i)\} \qquad d_A^- := \max\{d_A^-(i)\}$$

$$d_A := \max\{d_A^+, d_A^-\}$$

$$d^+ := \max\{d_{A_1}^+, d_{A_2}^+\} \qquad d^- := \max\{d_{A_1}^-, d_{A_2}^-\}$$

For the sake of completeness we also put a note on the case $m > n$:

Remark 3.3.2. It is always possible to interpret a double set cover problem as a network $G(V, E_1, E_2)$ with two arc sets, similar to the example depicted in 3.1 where we interpret E_1, E_2 as the catchment areas of the facilities. If $m \leq n$ the set $\mathcal{F}$ contains can be interpreted as a list of the neighbors, more precisely:

$$\mathcal{N}_{E_1}(i) = S_i^{(1)} \qquad \mathcal{N}_{E_2}(i) = S_i^{(2)}$$

If $\mathcal{F}$ defines a matrix that has $m > n$ we need to add auxiliary nodes in S and also elements

3.3. Complexity Results

in $\mathcal{F}$ to construct an equivalent $\mathcal{F}'$ that has a corresponding square matrices A'_1 and A'_2. Since $m > n$ the instance $\mathcal{F} = \{(\{2\}, \{2\}), (\{1,3\}, \{1\}), (\{1,2\}, \{1,2\}), (\{2,3\}, \{1,2\}), (\{1\}, \{3\}), (\{3\}, \{1,2\})\}$ suits as an example. The graph $G(V, E_1, E_2)$ of $\mathcal{F}$ is depicted in Figure 3.3. The shaded nodes denote $S = \{1,2,3\}$. The set to cover by $\mathcal{F}'$ is $S' = \{1, 2, 3, 4, 5, 6, 4', 5', 6', 4'', 5'', 6''\}$ and it augments S by $3(m-n) = 6$ new vertices. The vertices $\{4, 5, 6\}$ correspond to the overplus in $\mathcal{F}$, namely: $\{(\{2,3\}, \{1,2\}), (\{1\}, \{3\}), (\{3\}, \{1,2\})\}$. And since $\{4, 5, 6\}$ also need to be doubly covered we add $\{4', 5', 6'\}$ for service 1 and $\{4'', 5'', 6''\}$ for the service 2. The network on the left hand side corresponds to A_1 ($S_i^{(1)}$) and the one on right hand side corresponds to $A_2(S_i^{(2)})$. We can see that $\{4', 5', 6'\}$ as well as $\{4'', 5'', 6''\}$ have to be selected in Λ_1 and Λ_2, respectively. Therefore the corresponding double set cover problem $\mathcal{F}'$ plus defining $k' = k + 2(m-n)$ completes the transformation.

We will proof NP-completeness under low-key requirements to the density of A_1 and A_2, and we will use a reduction of the 3-SAT problem [18]:

Definition 3-SATISFIABILITY (3-SAT).
INSTANCE: Set U of variables, collection of clauses over U such that each clause $c \in C$ has at most 3 literals ($|c| \leq 3$) QUESTION: Is there a truth assignment for U that simultaneously satisfies all clauses?

Proposition 3.3.3. *3-SAT is NP-complete, even if every variable u appears in at most 5 different clauses, either as u or $\neg u$. [80]*

Remark 3.3.4. If $d^- \leq 2$ then we can interpret an instance of DSCP as a *Double Vertex Cover Problem*. First we will give the arguments for the first service $S_i^{(1)}$, since the arguments also apply for the second one. Focusing on the first service, every node in DSCP may be interpreted as an edge and the sets $S_i^{(1)}$ as vertices:

- If the node j is element of two sets $S_{i_1}^{(1)}$ and $S_{i_2}^{(1)}$, then j is an edge that connects the vertices $S_{i_1}^{(1)}$ and $S_{i_1}^{(1)}$. Therefore a vertex cover will contain at least one of them.

- If the node j is element of exactly one $S_i^{(1)}$ then j can be interpreted as a loop and the vertex $S_i^{(1)}$ has to be selected.

- If the node is not element of any $S_i^{(1)}$ then the instance is infeasible. Therefore we can neglect this case.

The same arguments hold for the second service $S_i^{(2)}$. Therefore we can conclude that the problem is a variant of the following *independent set vertex cover problem*:

3. The Double Set Cover Problem

Definition Independent Set Vertex Cover Problem(ISVCP). Given a graph $G(V_1, V_2, E)$ find a subset $\Lambda \subset V$. We define:
$$\Lambda_1 := \Lambda \cap V_2 \qquad \Lambda_1 := \Lambda \cap V_2$$
$$E_1 := \{[i,j] \in E : i, j \in V_1\} \qquad E_2 := \{[i,j] \in E : i, j \in V_2\}$$
and finally the bridges
$$E_{1,2} := \{[i,j] \in E : i \in V_1 \wedge j \in V_2\}$$
Then Λ has to have the following properties:

- Λ_1 is a vertex cover for $(V, E \setminus E_{12})$, or
    - Λ_1 is a vertex cover for (V_1, E_1).
    - Λ_2 is a vertex cover for (V_2, E_2).
- Λ is an independent set for (V, E_{12}).

Another problem that is related to the DSCP with $d^- \leq 2$ may be called:

Definition Peaceful Vertex Cover Problem (PVCP): Given a graph $G(V, E)$ and a conflict set $W \subset V \times V$. We want to find a subset $\Lambda \subset V$ such that:

- Λ is a vertex cover for (V, E).
- $(i,j) \in W \Rightarrow \neg(\{i,j\} \subset \Lambda)$.

Remark 3.3.5. Coming back to the the interpretation of the DSCP with $d^- \leq 2$ as IVCP we give a detailed description of the interpretation:

- $V_1 := \{v_i : S_i^{(1)} \neq \phi\}$
- $V_2 := \{w_i : S_i^{(2)} \neq \phi\}$
- $E_1 := \{e_1, \ldots, e_n\}$
- $E_2 := \{f_1, \ldots, f_n\}$
- $E_{1,2} := \{[v_i, w_i] : v_i \in V_1 \wedge w_i \in V_2\}$

If there are 2 possible sets $S_{i_1}^{(1)}$ and $S_{i_2}^{(1)}$ that may cover $j \in S$ then $f_j = [v_{i_1}, v_{i_2}]$. If there is only one possible set then $f_j = [v_i, v_i]$. E_2 is constructed analogously.

Theorem 3.3.6. *DSCP is NP-complete if $d_{A_1}, d_{A_1} \leq 2$.*

3.3. Complexity Results

Proof. Since 3-SAT is NP-complete also if , we can show NP-completeness by polynomially reducing 3-SAT (where each variable appears at most 5 times in a different clause) to DSCP. Suppose a number of variables $U = \{x_1, \ldots, x_n\}$ (x_i and $i \in \{1 \ldots n\}$) and a collection of clauses $C = \{c_1, \ldots, c_m\}$ (c_j and $j \in \{n+1 \ldots n+m\}$) with at most 3 literals is given. Some variables x_i are linked to some clauses c_j by (S_r^1, S_r^2) where $r = i_l = j_{\tilde{l}}$. I.e: we will construct an instance of DSCP with $d_{A_1} \leq 2$ and $d_{A_1} \leq 2$ and for the variables and clauses we construct circular gadgets regarding the first, and the second service. Using the analogy to the ISVCP the sets $S_i^{(1)}$ and $S_i^{(2)}$ can be interpreted as nodes and the set S can be interpreted as edges.

$$S := \bigcup_{i=1}^{n} \bigcup_{l=1}^{10} \{i_l\} \cup \bigcup_{j=n+1}^{m+n} \bigcup_{l=1}^{9} \{j_l\}$$

The key idea is, to represent each variables $x_i \in U$ as an even circle with 10 nodes $T_{i_l}^1$ ($l = 1 \ldots 10$) and to represent each clause $c_j \in C$ as an odd cycle with 9 nodes $T_{j_l}^2$ ($l = 1 \ldots 9$):

$$T_{i_l}^1 = \{i_l, i_{1+(l \bmod 10)}\}$$

$$T_{j_l}^2 = \{j_l, j_{1+(l \bmod 9)}\}$$

Suppose that the variables $\{x_{j_1}, x_{j_2}, x_{j_3}\}$ are contained in $c_j \in C$. For technical reasons, we define the rank rk_1 of a clause c_j for x_i and $\neg x_i$:

$$\text{rk}_1 : (C \cup \neg C) \times U \mapsto \{1 \ldots 10\}$$

$$\text{rk}_1(x_i, c_j) = |\{j' | j' \leq j \wedge x_i \in c_j \wedge x_i \in c_{j'}\}|$$

$$\text{rk}_1(\neg x_i, c_j) = |j' | j' \leq j \wedge \neg x_i \in c_j \wedge \neg x_i \in c_{j'}\}|$$

and we also define the rank rk_2 of literal in a clause:

$$\text{rk}_2 : \{1 \ldots n\} \times U \mapsto \{1 \ldots 3\}$$

$$\text{rk}_2(i, c_j) = |\{i' | i' \leq i \wedge x_i, \neg x_i, x_{i'}, \neg x_{i'} \in c_j\}|$$

Now we are ready to define $\mathcal{F}$:

- For each literal x_i ($\neg x_i$) in c_j we add the following pair of sets to $\mathcal{F}$:

$$x_i \in c_j \quad \Rightarrow \quad \mathcal{F} \leftarrow \mathcal{F} \cup \{(T_{i_{2\text{rk}_1(x_i,c_j)}}^1, T_{j_{3\text{rk}_2(x_i,c_j)-2}}^2)\}$$

3. The Double Set Cover Problem

$$\neg x_i \in c_j \quad \Rightarrow \quad \mathcal{F} \leftarrow \mathcal{F} \cup \{(T^1_{i_{2\text{rk}_1(x_i,c_j)-1}}, T^2_{j_{3\text{rk}_2(x_i,c_j)-2}})\}$$

- For those sets $T^1_{i_l}$, $T^2_{j_l}$ that have not been used in the last step we set:

$$\mathcal{F} \leftarrow \mathcal{F} \cup \{(T^1_{i_l}, \phi)\}$$

$$\mathcal{F} \leftarrow \mathcal{F} \cup \{(\phi, T^2_{j_l})\}$$

- To complete the instance we have to add:

$$\forall_{i=1}^{n} \forall_{l=1}^{10} \quad \mathcal{F} \leftarrow \mathcal{F} \cup \{(\phi, \{i_l, i_{1+(l \bmod 10)}\})\}$$

$$\forall_{i=n+1}^{n+m} \forall_{l=1}^{9} \quad \mathcal{F} \leftarrow \mathcal{F} \cup \{(\{j_l, j_{1+(l \bmod 9)}\}, \phi)\}$$

Since the mentioned odd cycles and mentioned even cycles need at least 5 nodes to be covered (for each service type) we can conclude that the absolute minimum number to cover all circles is $10(|U| + |C|)$. Now we will show that this is possible to achieve if and only if the clauses C are satisfiable. To illustrate the construction we give an example for the clause $c_j = \{\neg x_1, \neg x_2, x_3\}$. The circular configuration for x_i and c_j are depicted on the left and respectively on the right side of figure 3.4. The linkage between these items is illustrated in figure 3.5. Since x_1 is negated in the clause an odd node of it's configuration is used. For x_3 which is not negated an even node is used. The circles for x_i are connected with the circle for the clause in a way that there are always 2 nodes left between. Each circle for a x_i has two minimal configurations, called the "true"- and the "false"- configuration. If all the odd nodes are selected then we say "$x_i =$ true", else "x_i=false". Now we show that it is possible to find a covering that uses only 5 sets if c_j =true, whereas if c_j =false it's not. we investigate the following assignments according to the example:

- If c_j =false then all literals are false, therefore $x_1 = x_2$=true and x_3=false. That means that $T^2_{j_1}$, $T^2_{j_4}$ and $T^2_{j_7}$ cannot be selected. Therefore we can only reach a covering with 6 sets. Figure 3.6 shows this situation - the nodes that appear together in $\mathcal{F}$ are the elements that are combined in the same dashed ellipse.

- If c_j =true then at least one of the literals is true: i.e.: x_1 =false, x_2 =false or x_3 =true. therefore we have three cases:

  ⊢ If all literals are true, then we are free to choose, therefore it we can realize any covering that uses 5 sets.

3.3. Complexity Results

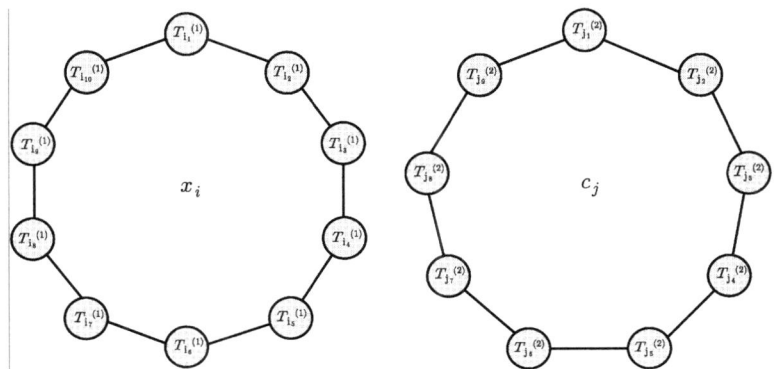

Figure 3.4.: Transformation of literal i and clause j into circular configurations

- If two literals are true, then we cannot choose one of the sets and it's neighbors have to be chosen. We can start on one of the sides and then we leave out every second note until we reach a feasible solution that uses 5 sets.
- If only one literal is true, then we cannot choose two of the sets. We notice that there are 2 sets between them, and both have to be chosen. We again start selecting on one of the sides and we leave out every second. This situation is depicted in Figure 3.7.

If C is satisfiable we directly get a solution with the minimum number of $10\,(|U|+|C|)$. On the other hand: If we get a minimal solution of the derived problem that only chooses $10\,(|U|+|C|)$ sets, then the sets in the even circles (variables) are alternately selected and deselected (x_i is "true" or "false"). Furthermore each odd circle (clause), there is at least one of the positions $1, 4$ or 7 that has to be deselected. Otherwise we have to choose all the sets between them - and that means 6 sets. So, one of the sets is selected and therefore the corresponding clause has a certain value, "true" or "false". In other words, each clause realizes at least one literal and each variable can only hold either the value "true" or "false". Therefore by extracting the values of the even circles we get a solution to the original problem.

□

Corollary 3.3.7. *As a direct consequence the ISVCP is NP-complete, also if E_1 and E_2 only consist of disjoint paths and circles and where $E_{1,2}$ consists of disjoint edges.*

Corollary 3.3.8. *Similarly we can see that the PVCP is NP-complete, also if E only consist of disjoint paths and circles and if for each node there exists at most one conflict.*

3. The Double Set Cover Problem

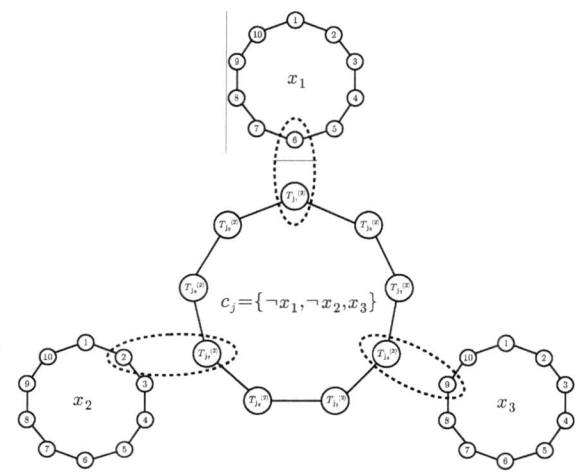

Figure 3.5.: Example of a clause c_j in tree variables x_1 x_2 x_3

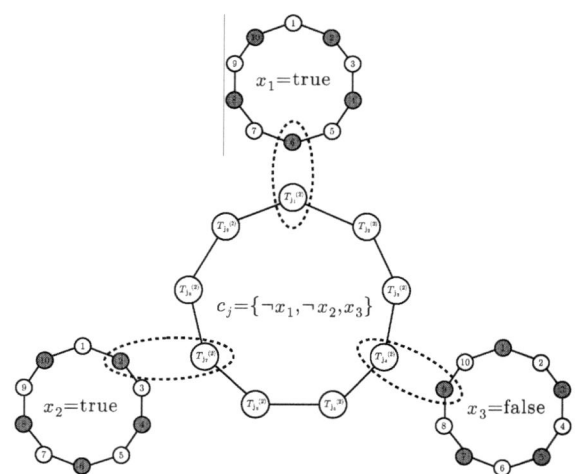

Figure 3.6.: Example where the clause c_j=false

3.3. Complexity Results

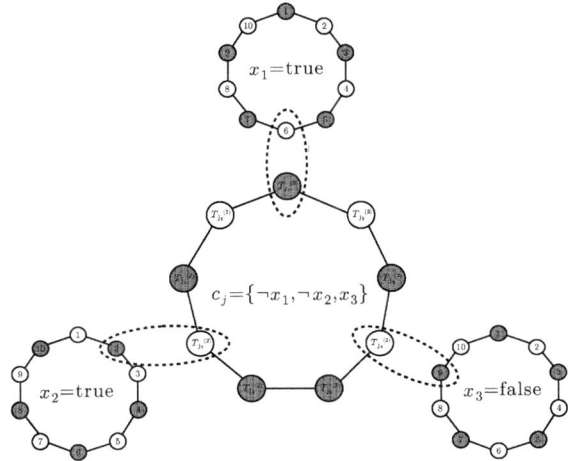

Figure 3.7.: Example where the clause c_j=true

In the next investigations we will use a property that guarantees the existence of integral solutions for systems of linear equations. We present the definition of total unimodularity and a sufficient criterion.

Definition *totally unimodular* (TUM): A given matrix is called total unimodular if the determinants of all square submatrices (minors) are element of $\{-1, 0, 1\}$.

Proposition 3.3.9. *A matrix A is totally unimodular if and only if A^t is totally unimodular.*

Proof. The proof is a direct consequence of $\det(B) = \det(B^t)$. □

Proposition 3.3.10. *Integer programming, which is known to be NP-hard gets easy to solve (in polynomial time) if the coefficient matrix is total unimodular. (see [59] and [66])*

Proposition 3.3.11. *Sufficient criterion for* total unimodularity *[57]:*
A matrix that only contains zeros and ones is totally unimodular if it is possible to partition the matrix into two sets of rows such that each column has one entry in each of these sets.

Proof. Suppose that a matrix A that fulfills the requirements is given and let's suppose that the statement is not true. The partition (I_1, I_2) is given and it is clear that all minors of $(1 \times 1)-$submatrices are 0 or 1. Now suppose that r is the largest value where all minors of $(r \times r)-$submatrices of B are element of $\{-1, 0, 1\}$. Therefore $1 \leq r < n$ and there exists a submatrix

39

3. The Double Set Cover Problem

B whose determinant is not element of $\{-1, 0, 1\}$. We will show that B has exactly two entries in each column. Because if not, then there exists a column with no entry or only one entry.

- no entry implies that we have a column of zeros and therefore $\det(B) = 0$.

- if there is only one entry, then we can use Laplace's Formula to expand the calculation of the determinant along this column. Since we only have one summand the determinant is $\pm$ a minor of a $(r-1 \times r-1)$-matrix and therefore $\det(B) \in \{0, \pm 1\}$

Both cases are irrelevant and we conclude that all columns have 2 entries. Now we can separate the rows $j \in J$ of $B = (b_j^t)$ according to the partition (I_1, I_2) and sum them up.

$$\sum_{j \in I_1 \cap J} b_j = \sum_{j \in I_1 \cap J} b_j$$

That shows that the rows of B are linearly dependent and therefore $\det(B) = 0$ - a contradiction. Now we show that there exist matrices that are TUM and don't show don't satisfy the properties of the criterion.

$$A = \begin{pmatrix} 1 & 1 & 0 & 0 & 0 \\ 1 & 0 & 1 & 0 & 0 \\ 1 & 0 & 0 & 1 & 1 \end{pmatrix}$$

The given matrix A is TUM and since there are columns that have 3 entries we cannot make a partition that has the wanted property (also for the transposed). □

Theorem 3.3.12. *If $d \leq 2$ and $m = n$ then DSCP is polynomial.*

Proof. Again, lets suppose that $d_{A_1} = d_{A_1} = 2$ then A_1 and A_2 may be interpreted as singletons and edges that have at most one vertex in common. For each service we have to cover n nodes. Since $n = m$ we can interpret the instance as mentioned in remark (3.3.2) - each vertex $i \in S$ covers at most 2 vertices in S. Since we have to reach each node we have:

$$2|\Lambda_1| \geq |S| \quad \wedge \quad 2|\Lambda_2| \geq |S|$$

$$\Rightarrow |\Lambda_1| + |\Lambda_2| \geq |S|$$

therefore we need to select all vertices, i.e.: $\Lambda_1 \cup \Lambda_2 = S$. That also means that each vertex has to cover exactly 2 vertices. If a vertex i covers less then 2 vertices in the first service, then it has to cover 2 vertices in the other service, i.e.: $i \in \Lambda_2$. Therefore we suppose that these decisions are already fixed and we only have to decide upon those nodes where we have both options. By eliminating all these options (recursively) we may end up in having no choice left (infeasibility or feasibility) or we can reduce the matrices A_1 and A_2 to the matrices $\widetilde{A_1}$ and $\widetilde{A_2}$

that have exactly 2 row entries and exactly 2 column entries. Now, we are ready to reformulate the following problem:

$$\widetilde{A_1}x = 1 \tag{3.22}$$
$$\widetilde{A_2}y = 1 \tag{3.23}$$
$$x + y = 1 \tag{3.24}$$
$$x_i, y_i \in \{0, 1\} \tag{3.25}$$

or to put it different:

$$A \begin{pmatrix} x \\ y \end{pmatrix} = \begin{pmatrix} \widetilde{A_1} & \\ & \widetilde{A_2} \\ I_n & I_n \end{pmatrix} \begin{pmatrix} x \\ y \end{pmatrix} = 1$$

Now we will show that this coefficient matrix is total unimodular (TUM) and therefore the problem is solvable in polynomial time. In fact any basic solution of the system gives us a feasible integer solution. We have to recall that $\widetilde{A_1}$ and $\widetilde{A_2}$ have exactly 2 entries in each row and column and therefore we can represent them as disjoint circles. If one of the circles is odd then the problem is infeasible. Now, lets suppose that all circles are even, then it is possible to give a bipartition (Γ_1^1, Γ_2^1), (Γ_1^2, Γ_2^2) of the nodes $\widetilde{V}$, for $\widetilde{A_1}$ and $\widetilde{A_2}$ respectively. Applying the criterion for total unimodularity (Theorem 3.3.11) on A^t with the index sets $(\Gamma_1^1 \cup \Gamma_1^2, \Gamma_2^1 \cup \Gamma_2^2)$ we get the result. □

A practical method to solve the problem is a backtracking method: i.e.: set $x_1 = 1$ and fix all variables that appear in S_i^1 and S_i^2 accordingly. Since x and y are alternatives we only have to fix the values for x. We continue this process of fixing x until we reach a contradiction or alternatively can't find a new neighbor. The pseudo code of the algorithm is given in Algorithm 1. The algorithm uses the sets L and R that represent possible conflicts. I.e. if $(i, 1)$ is element of R then x_1 should take the value 1. Therefore, if the Algorithm 1 terminates with v=feasible then L gives us the opposite values of the correct variables. I.e.:

$$(i, l) \in L \Rightarrow x_i \leftarrow (1 - l)$$

. To estimate the complexity of the Algorithm we have to notice that starting in a small circle might lead to a bad performance: The algorithm runs in $O(n^2)$ time.

Remark 3.3.13. The key argument of Theorem 3.3.12 is that A_1 and A_2 can be interpreted as disjoint even circles, which directly leads us to a suitable row partition.

3. *The Double Set Cover Problem*

Algorithm 1 A backtracking method

$L \leftarrow \phi$;
$v \leftarrow$ feasible;
while $(|L| < |S|)$ **do**
    $i \leftarrow \min\left(\{t \in S : (t,0) \notin L \wedge (t,1) \notin L\}\right)$;
    $L' \leftarrow \{(i,0)\}$;
    $R \leftarrow \{(j,1) : j \in S_i^1\} \cup \{(j,0) : j \in S_i^2\}$;
    while $(R \neq \phi)$ **do**
        $r \leftarrow (j,l) \in R$;
        if $(r \notin L \cup L')$ **then**
            $R \leftarrow R \setminus \{(j,l)\}$;
            $L' \leftarrow L' \cup \{(j, 1-l)\}$;
            $R \leftarrow R \cup \{(s,l) : s \in S_j^1\} \cup \{(s, 1-l) : s \in S_j^2\}$;
        else
            if $((i,1) \in L')$ **then**
                $R \leftarrow \phi$;
                $L \leftarrow \{(s,1) : s \in S\}$;
                $v \leftarrow$ infeasible;
            else
                $L' \leftarrow \{(i,1)\}$;
                $R \leftarrow \{(j,0) : j \in S_i^1\} \cup \{(j,1) : j \in S_i^2\}$;
            end if
        end if
    end while
    if $(v =$ feasible$)$ **then**
        $L \leftarrow L \cup L'$;
    end if
end while
return v;

3.3. Complexity Results

If an instance satisfies $d^+ \leq 2$ it is possible to interpret the DSCP as an edge covering problem and incites us to formulate the following problem:

Definition Peaceful Perfect Matching Problem (PPMP): Given a graph $G(V, E)$ and a conflict set $W \subset E \times E$. We want to find a subset $\Lambda \subset V$ such that:

- M is a perfect matching of (V, E).

- $(e_i, e_j) \in W \Rightarrow \neg(\{e_i, e_j\} \subset M)$.

Since the odd cycles in Theorem 3.3.6 are essential we needed to choose $k > n$ sets from $\mathcal{F}$. If $(k = n)$ we can conclude that we have chosen perfect matchings for both services, and therefore disjoint odd cycles cannot be present. Like mentioned in Remark 3.3.13 even cycles make the problem easier and we can also show that the following statement is true:

Corollary 3.3.14. *If $d_{A_1}, d_{A_2} \leq 2$ and $k = n$ then DSCP is polynomial.*

Proof. Similarly to the proof for Theorem 3.3.6 we can show that $d_{A_1}, d_{A_2} \leq 2 \Rightarrow k \geq |\Lambda_1 \cup \Lambda_2| \geq n$. Therefore $k = n$ is the minimum we can reach and we can concentrate on selecting edges. We again reduce the matrices A_1 A_2 and build up an auxiliary linear program that solves the problem. Like in the proof of Theorem 3.3.6 we filter out the real alternatives $\widetilde{A_1^1}, \widetilde{A_2^1}$ and we additionally integrate those edges that may only be selected for one of the services $(\widetilde{A_1^2}, \widetilde{A_2^2})$. We emphasize that we already exclude paths from our investigations, since they don't leave a choice. We state the following linear programm:

$$\min \sum_{i=1}^{m_1} (x_i^1 + y_i^1) + \sum_{i=1}^{m_2} (x_i^2 + y_i^2) \qquad (3.26)$$

$$\widetilde{A_1^1} x^1 + \widetilde{A_1^2} x^2 = 1 \qquad (3.27)$$

$$\widetilde{A_2^1} y^1 + \widetilde{A_2^2} y^2 = 1 \qquad (3.28)$$

$$x^1 + y^1 \leq 1 \qquad (3.29)$$

$$x_i^1, x_i^2, y_i^1, y_i^2 \in \{0, 1\} \qquad (3.30)$$

or to put it different:

$$\left| A \begin{pmatrix} x^1 \\ y^1 \\ x^2 \\ y^2 \end{pmatrix} \right| = A = \begin{pmatrix} \widetilde{A_1^1} & & \widetilde{A_1^2} & \\ & \widetilde{A_2^1} & & \widetilde{A_2^2} \\ I_{m_1} & I_{m_1} & & \end{pmatrix} \begin{pmatrix} x^1 \\ y^1 \\ x^2 \\ y^2 \end{pmatrix} \overset{\leq}{\underset{=}{}} 1$$

3. The Double Set Cover Problem

The matrix A of the coefficients includes $(\widetilde{A_1^1}, \widetilde{A_1^2})$ which represents a set of disjoint even cycles. The same is true for $(\widetilde{A_2^1}, \widetilde{A_2^2})$, therefore we may partition the columns in a way that suits to criterion 3.3.11, and we again find total unimodularity. □

Now, we give an alternative proof of NP-completeness where $k = n$ and $d_{A_2} \leq 2$ and for the other service we take $d_{A_1}^{(-)} \leq 2$ and $d_{A_1}^{(+)} = 3$. First, we will show for PPMP the following:

Proposition 3.3.15. *Complexity of the Peaceful Perfect Matching Problem:*

1. *For $d(i) \leq 2$ PPMP is polynomially solvable*

2. *If $d(i) \leq 3$ then PPMP is NP-complete*

Proof. The proof has two parts,

1. It is clear that E has to consist of disjoint paths of odd length and circles of even length. Considering feasibility, we need to take the conflict set W into account. To simplify the problem we represent each circle and each path with a literal $x_i (i \in N)$. Since there are only 2 possible configurations we assign the value $x_i = 0$ for one of the configurations and $x_i = 1$ for the other. Each conflict can be translated: i.e.: if $(i, j) \in W$ then we have two possible situations:

    - i and j are members of the same circle, then obviously the conflict is either indissoluble or dispensable.

    - i and j are members of different circles x_i and x_j, then the conflict translates into two conflicting situations $(a_i, a_j \in \{0, 1\})$. That means that it is not possible to have $x_i = a_i$ and $x_j = a_j$ at the same time. Since $x_i \neq a_i \Rightarrow x_i + a_i = 1$ the conflict can be written as the following linear equation:

    $$(2a_i - 1)(x_i + a_i - 1) + (2a_j - 1)(x_j + a_j - 1) \leq 1$$

    To solve this system of linear equations we can adopt the Algorithm 1 accordingly. A pseudo code to solve linear inequalities mentioned above is given in Algorithm 2.

2. Now we will show that that PPMP is NP-complete by using a reduction of 3-SAT. Since the proof is similar to Theorem 3.3.6 we can reduce the details and only give a sketch of the proof: Roughly speaking, we are searching for non-conflicting perfect matchings. For each clause we define squares isomorphic to K_4 and for each variable we construct a circle that is sufficiently large. Each of the matchings of the squares represent a literal. I.e.: Each one of the 3 possible matchings on the square selects which literal has to

3.3. Complexity Results

Algorithm 2 Solving a system of linear inequalities with two binary variables per inequality

$L \leftarrow \phi$;
$v \leftarrow$ feasible;
while ($|L| < |N|$) **do**
    $i \leftarrow \min(\{i \in N : x_i = 0 \notin L \wedge x_i = 1 \notin L\})$;
    $L' \leftarrow \{x_i = 0\}$;
    $R \leftarrow \{x_j = a_j : ((1 - x_i) + (2a_j - 1)(x_j + a_j - 1) \leq 1) \in LP\}$;
    while ($R \neq \phi$) **do**
        $r \leftarrow x_j = a_j \in R$;
        if ($r \notin L \cup L'$) **then**
            $R \leftarrow R \setminus \{r\}$;
            $L' \leftarrow L' \cup \{r\}$;
            $R \leftarrow R \cup \{x_s = a_s : ((2a_j - 1)(x_j + a_j - 1) + (2a_s - 1)(x_i + a_s - 1) \leq 1) \in LP\}$;
        else
            if ($x_i = 1 \in L'$) **then**
                $R \leftarrow \phi$;
                $L \leftarrow \{x_s = 0 : s \in N\}$;
                $v \leftarrow$ infeasible;
            else
                $L' \leftarrow \{x_i = 1\}$;
                $R \leftarrow \{x_s = a_s : (x_i + (2a_s - 1)(x_s + a_s - 1) \leq 1) \in LP\}$;
            end if
        end if
    end while
    if ($v =$ feasible) **then**
        $L \leftarrow L \cup L'$;
    end if
end while
return v;

3. The Double Set Cover Problem

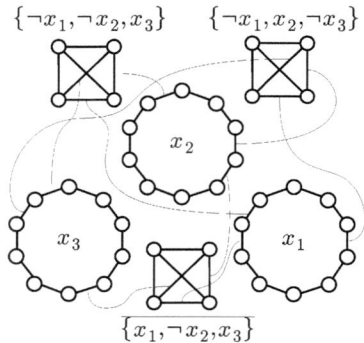

Figure 3.8.: Example of a transformation of a 3-SAT instance with 3 clauses and 3 variables

be fulfilled. We select 3 distinguished edges in the square (c_j) that belong to different matchings and indicate the corresponding literals $l_{j_1}, l_{j_2}, l_{j_3}$. For each even length circle that represents a variable (x_i) we exactly have two possible configurations - one of them indicates that x_l =true and the other one x_l =false. According to that we have even and odd edges in each circle. If the even edges are chosen to be part of the matching, then x_i =true therefore the edges are also called to be "true", analogously the odd edges are called "true". For each clause $c_j = \{l_{j_1}, l_{j_2}, l_{j_3}\}$ we construct conflicts between squares and circle. We illustrate the transformation by giving an example. Figure 3.8 gives a transformation of an instance and figure 3.9 gives a solution.

□

Corollary 3.3.16. *If $k = n$, $d_{A_1}^{(+)}, d_{A_1}^{(-)} \leq 2$ and $d_{A_2}^+ \leq 2, d_{A_2}^- \leq 3$ then DSCP is NP-complete.*

Proof. We can interpret the instance constructed in the second part of proof of theorem 3.3.15 as an instance of DSCP. The idea is to interpret the edges that build the circles as part of A_1 and the edges that are part of the squares as part of A_2. That means if the number of nodes from squares is larger than the number of nodes from circles, then the number of elements in S is equal to the number of nodes from the squares else we take the circles. If we find a conflict between two edges then the corresponding sets are joined in one pair of $\mathcal{F}$. Since some elements cannot be covered we complete the instance with adequate sets. We illustrate this transformation in Figure 3.10. □

Lemma 3.3.17. *DSCP is polynomial if $A_1 = A_2 = A$ and $d_A \leq 2$.*

3.3. Complexity Results

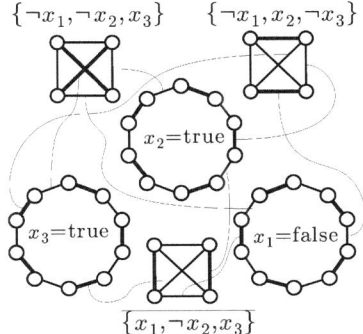

Figure 3.9.: Solution to the example given in Figure 3.8

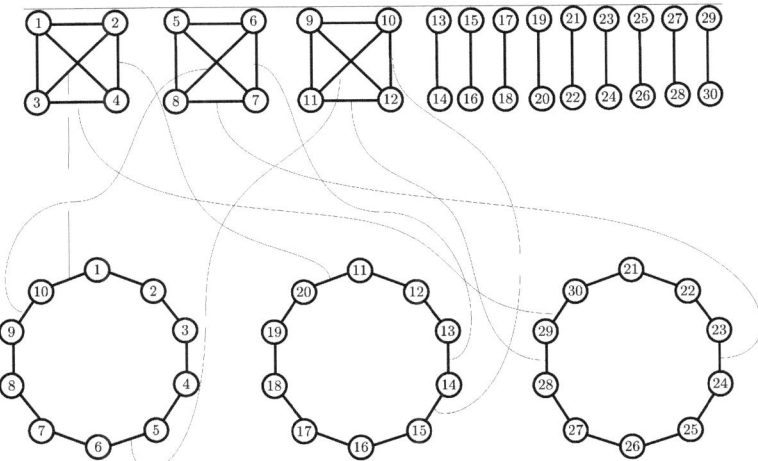

Figure 3.10.: Example of a transformation of PPMP to DSCP to apply Proposition 3.3.15

47

3. The Double Set Cover Problem

Proof. We can only find a small number of different situations:

- even circles: Suppose that $v_i \in \{1\ldots 2l\} \subset S$ are the elements of the circle C that are connected by $S_i^1 = S_i^2 = S_i = \{v_i, v_{1+(i \bmod 2l)}\}$. Then the circle $C = \bigcup_{i=1}^{2l} \{(S_i, S_i)\}$ can be handled by choosing alternately the $(2i)$ for Λ_1 and $(2i-1)$ for Λ_2.

- odd circles $\bigcup_{i=1}^{2l+1} \{(\{v_i, v_{1+(i \bmod 2l)}\}, \{v_i, v_{1+(i \bmod 2l)}\})\}$ lead to infeasibility.

- odd path $\bigcup_{i=1}^{2l+1} \{(\{v_i, v_{i+1}\}, \{v_i, v_{i+1}\})\}$ can be handled by alternately choosing the sets for Λ_1 and Λ_2.

- even paths $\bigcup_{i=1}^{2l} \{(\{v_i, v_{i+1}\}, \{v_i, v_{i+1}\})\}$ make the instance infeasible.

- singletons $f = (\{v_i\}, \{v_i\}) \in \mathcal{F}$ could be found isolated or at the endings of a path, which leads back to odd and even paths.

□

Lemma 3.3.17 shows that $A_1 \neq A_2$ is necessary to prove NP-completeness and gives reason to investigate the case $A_1 = A_2$. Now, we will show that DSCP is NP-complete if $A_1 = A_2 = A$, $d_A^+ \leq 2$ and $d_A^- \leq 3$.

To continue discussing the case $A_1 = A_2$ we consider the case where $d^+ \leq 2$ and $d^- \leq 3$. We suppose that $d^+ = 2$ and we can interpret $\mathcal{F}$ as a collection of edges. As a further restriction we concentrate on instances where $k = n$. That is equivalent to the question if we can find two disjoint perfect matchings in $\mathcal{F}$. If $\mathcal{F}$ represents a cubic graph then it is equivalent to finding a 1-factorization of the graph. In other words if we can show that it is hard to find a 1-factorization of a cubic graphs then we are done [100]. CHROMATIC-INDEX for Edge Coloring is equivalent factorizing a graph into 1-factors. This problem is known to be NP-hard and [61] shows that CHROMATIC-INDEX is also NP-hard for cubic graph. Therefore we state the following result:

Corollary 3.3.18. *DSCP is NP-complete even if $A_1 = A_2$ and $d^+ \leq 2$ and $d^- \leq 3$.*

As a last step in analyzing the case $A_1 = A_2$ we consider the case where $d^- \leq 2$.

Theorem 3.3.19. *DSCP is polynomially solveable if $A_1 = A_2$ and $d^- \leq 2$.*

Proof. We can interpret DSCP to a 2-coloring vertices which is easy to solve: Since $d^- \leq 2$. for each $i \in S$ there exist at most two sets $S_{i_1}, S_{i_2} \in \mathcal{F}$. We can find 3 situations:

3.3. Complexity Results

1. If $d^-(i) = 0$ then DSCP is infeasible since there is no set that covers i.
2. If $d^-(i) = 1$ then there is only one set that covers i, therefore can be either selected for Λ_1 or Λ_2. Therfore the instance is infeasible.
3. If $d^-(i) = 2$ we can think of $i = [S_{i_1}, S_{i_2}]$ as an edge and S_j build the nodes.

Only the last case is of interest, therefore we may interpret the problem as coloring of the vertices: Suppose that there is an edge that in the auxiliary graph where both endpoints are chosen for the same service, then this edge is isolated from the other service and we end up in infeasibility. But checking if a graph is 2-colorable is an easy task. I.e.: start with an arbitrary node and assign one of the colors. In the next steps we assign an appropriate color to all new neighbors that are not yet colored and those that are already colored are checked for feasibility. In this process each edge of the auxiliary graph is visited exactly once - therefore we end up in a algorithm with $O(n)$. □

We summarize the complexity results for the DSCP in the graph depicted in Figure 3.11. We starts with the unrestricted version on top and successively add restrictions. Every path classifies a sub problem of DSCP as NP-complete or polynomially solvable.

3.3.2. Complexity Results for the Optimization Versions of the DSCP

Regarding the corresponding optimization version we may decide for a DSCP instance if we can find a proper solution or not, i.e:

Corollary 3.3.20. *ODSCP1 and ODSCP2 are NP-hard, even if $d(A_1), d(A_2) \leq 2$.*

Proof. It is easy to see that an instance $\mathcal{F}$ of DSCP is feasible if and only if the objective value of the corresponding optimization versions is zero ($f^*_{ODSCP1} = 0$ and $f^*_{ODSCP1} = 0$). Therefore it is also possible to make a transfer of the assumptions of the Theorem 3.3.6. □

Lemma 3.3.21. *ODSCP1 is polynomial if $A_1 = A_2 = A$ and $d_A \leq 2$.*

Proof. This proof is done by inspection. □

Lemma 3.3.22. *For ODSCP1 is possible to reduce instances that satisfy $d(A_1)^+ \leq c$ to equivalent instances with $d(\widetilde{A_1})^+ = c$.*

Proof. I.e: We will construct an instance mainly by adding $c + 2$ auxiliary sets $\left\{ \left(\widetilde{S^1_{w_1}}, \widetilde{S^2_{w_1}} \right), \right.$ $\left. (\widetilde{S^1_{w_2}}, \widetilde{S^2_{w_2}}), \ldots (\widetilde{S^1_{w_{c+2}}}, \widetilde{S^2_{w_{c+2}}}) \right\}$ to $\mathcal{F}$ and the following definitions:

3. The Double Set Cover Problem

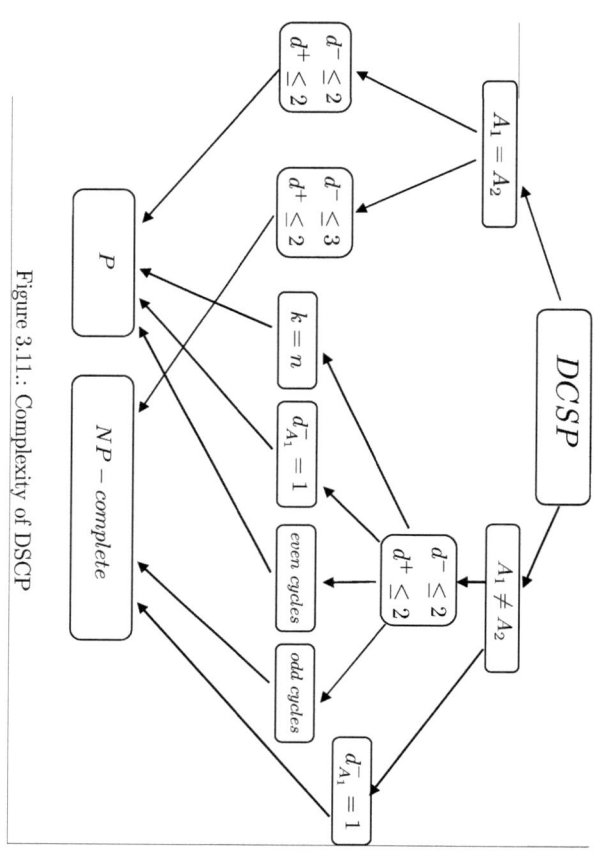

Figure 3.11.: Complexity of DSCP

3.3. Complexity Results

- $\tilde{k} = k + c + 2$
- $\widetilde{S_i^2} = S_i^2$
- $\widetilde{S_i^1} = S_i^1 \cup \{w_i : i \leq c - |S_i|\}$
- $\widetilde{S_{w_{c+1}}^1} = \{w_1, \ldots w_{c-1}\} \cup \{w_{c+2}\}$
- $\widetilde{S_{w_{c+2}}^1} = \{w_2, \ldots w_c\} \cup \{w_{c+1}\}$
- $\widetilde{S_{w_i}^1} = \{w_1, \ldots w_{c+1}\} \setminus \{w_i\}$ for $(i \leq c)$

On the other hand we need to cover the auxiliary nodes by the second service. Therefore we define the sets $\widetilde{S_{w_i}^2}$:

- $\widetilde{S_{w_i}^2} = \{w_{1+(i \bmod (c+2))}\}$ for $(i \leq c + 2)$

Now, every subset $\widetilde{S_i^1}$ has exactly c elements and obviously, w_{c+1} has to be selected ($w_{c+1} \in \widetilde{\Lambda_1} \cup \widetilde{\Lambda_2}$, where $\widetilde{\Lambda_1}$ $\widetilde{\Lambda_2}$ is the optimal solution of the derived instance), since $S_{w_{c+1}}^1$ is the only set that covers w_{c+2}. Therefore we only need to put effort on covering the remaining elements $\{w_c, w_{c+1}\}$. A good choice is selecting w_{c+2}, since it covers both and w_{c+1} is only covered by auxiliary sets $\widetilde{S_{w_i}^1}$. We conclude that we have to select at least 2 auxiliary elements and choosing $\{w_{c+1}, w_{c+2}\} \subset \Lambda_2$ is optimal, because of the structure of $\widetilde{S_{w_i}^2}$. I.e: regarding the second service, we need to select all auxiliary elements $\{w_1, \ldots w_{c+2}\} \subset \Lambda_1' \cup \Lambda_2'$ and there exists an optimal solution that satisfies:

- $\{w_{c+1}, w_{c+2}\} \subset \widetilde{\Lambda_1}$
- $\{w_1, \ldots w_c\} \subset \widetilde{\Lambda_2}$

We summarize, that we can transfer an optimal solution of the derived problem to the original one by setting:

- $\Lambda_1 = \widetilde{\Lambda_1} \setminus \{w_{c+1}, w_{c+2}\}$
- $\Lambda_2 = \widetilde{\Lambda_2} \setminus \{w_1, \ldots w_c\}$

Obviously, the other way round is also possible, and an optimal solution of the original problem leads to an optimal solution of the derived one:

- $\widetilde{\Lambda_1} = \Lambda_1 \cup \{w_{c+1}, w_{c+2}\}$
- $\widetilde{\Lambda_2} = \Lambda_2 \cup \{w_1, \ldots w_c\}$

3. The Double Set Cover Problem

That shows that the problems are equivalent and that the structure of the derived problem regarding the second service is minimally changed. I.e.: If the original instance is feasible, then $d(\widetilde{A_2})^+ = d(\widetilde{A_2})^+$. □

Remark 3.3.23. Because of the symmetry of the problem, Lemma 3.3.22 can also be applied for A_2 and therefor shows that it is possible to reduce instances that satisfy $d(A_2)^+ \leq c$ to equivalent instances with $d(A_2)^+ = c$.

To simplify some proofs, we will give conditions where ODSCP1 and ODSCP2 merge to one and the same problem:

Lemma 3.3.24. *If $d(A_1)^+(i) + d(A_2)^+(i) \leq 3$, then ODSCP1 and ODSCP2 may be regarded as equivalent.*

Proof. Suppose that (Λ_1, Λ_2) is optimal for ODSCP1 and the objective $f_{ODSCP1} > 0$, then we will investigate the elements $l_1 \in L_1 \subset \Lambda_1$ and $l_2 \in L_2 \subset \Lambda_2$ $(L_1 \cup L_2 \neq \phi)$ where we define consider the following sets $J^1(l) \subset J^1$ and $J^2(l) \subset J^2$:

$$J^1 = \{j : \sum_{i \in S} a_{ij}^{(1)} z_i^{(1)} = 0\}$$

$$J^1(l) = \{j : \sum_{i \in S} a_{ij}^{(1)} z_i^{(1)} = 0 \wedge a_{lj}^{(1)} = 1\} \quad l \in L_2$$

$$J^2 = \{j : \sum_{i \in S} a_{ij}^{(2)} z_i^{(2)} = 0\}$$

$$J^2(l) = \{j : \sum_{i \in S} a_{ij}^{(2)} z_i^{(2)} = 0 \wedge a_{lj}^{(2)} = 1\} \quad l \in L_1$$

$$f_{ODSCP1} = |J^1| \cup |J^2|$$

we can formulate the auxiliary SCP subproblems:

$$\min |\Delta_1| \quad \bigcup_{l \in \Delta_1} J^2(l) = J^2 \quad \Delta_1 \subset L_1$$

$$\min |\Delta_2| \quad \bigcup_{l \in \Delta_2} J^1(l) = J^1 \quad \Delta_2 \subset L_2$$

Since $i \in L : z_i^1 = 1 \Rightarrow d(A_1)^+(i) \geq 1 \Rightarrow d(A_2)^+(i) \geq 2$ this problem is easy to solve, since it is equivalent to the edge cover problem (using Lemma 3.3.22). The derived solution is:

$$\Lambda_1 \leftarrow \Lambda_1 \cup \Delta_2$$

3.3. Complexity Results

$$\Lambda_2 \leftarrow \Lambda_2 \cup \Delta_1$$

Since $f_{ODSCP2} = |\Delta_1| + |\Delta_2|$, this solution is feasible for ODSCP2 with $k \leftarrow k + |\Delta_1| + |\Delta_2|$ and $f_{ODSCP2} \leq f_{ODSCP2}$. On the other hand, taking an optimal solution of ODSCP2 we can generate a corresponding feasible solution for ODSCP1 by investigating $\Lambda_1 \cap \Lambda_2$. First we define:

$$\widetilde{\Lambda_1} = \Lambda_1 \setminus (\Lambda_1 \cap \Lambda_2)$$

$$\widetilde{\Lambda_2} = \Lambda_2 \setminus (\Lambda_1 \cap \Lambda_2)$$

Since,

$$i \in \Lambda_1 \cap \Lambda_2 \Rightarrow d(A_1)^+(i) = 1 \vee d(A_2)^+(i) = 1$$

we can separate $\Lambda_1 \cap \Lambda_2$ into disjoint sets Δ_1, Δ_2:

$$\Delta_1: \quad i \in \Delta_1 \subset (\Lambda_1 \cap \Lambda_2) \Leftrightarrow d(A_1)^+(i) = 2$$

$$\Delta_2: \quad i \in \Delta_2 \subset (\Lambda_1 \cap \Lambda_2) \Leftrightarrow d(A_2)^+(i) \leq 1$$

Now, we define the solution for ODSCP2:

$$\widetilde{\Lambda_1} \leftarrow \widetilde{\Lambda_1} \cup \Delta_1$$

$$\widetilde{\Lambda_2} \leftarrow \widetilde{\Lambda_2} \cup \Delta_2$$

Additionally, we know that for every $i \in \Delta_1$ and $j \in J^2$ we have:

$$\sum_{i \in \Delta_1} a_{ij}^{(2)} = 1 \quad \wedge \quad \sum_{j \in J^2} a_{ij}^{(2)} = 1$$

Else the solution for ODSCP2 could be improved by eliminating i from Λ_2. Obviously, k is decreased by $|\Lambda_1 \cap \Lambda_2|$, i.e:

$$k \leftarrow k - |\Lambda_1 \cap \Lambda_2|$$

Summarizing, it follows that $|\Delta_1| = |J^2|$ and $|\Delta_2| = |J^1|$, or $f_{ODSCP1} = f_{ODSCP2}$, which completes the proof. □

Lemma 3.3.25. *ODSCP1 and ODSCP2 are polynomial if $d_{A_1} = 1$ and $d_{A_2}^+ \leq 2$.*

Proof. Ad ODSCP1: Since $d_{A_1} = 1$ we have to select all vertices, i.e.: $\Lambda_1 \cup \Lambda_2 = S$. If $\Lambda_1 = S$ and $\Lambda_2 = \phi$ we have an error of size $|S|$. Now we suppose that $\Lambda_2 \neq \phi$, then Λ_2 defines a set of edges, singletons and empty sets. Since Λ_1 and Λ_2 are disjoint sets and all nodes in S are

3. The Double Set Cover Problem

covered by exactly one set S_i^1 the objective value is only dependent on Λ_2 or on how to cover S with the minimum number of sets from $\mathcal{F}^\in = \{S_i^2\}$. We will argue that it is possible to rewrite the objective function in the following way:

$$f^* = \min f(\Lambda_2) = |\Lambda_2|$$

To proof this, we suppose that (Λ_1, Λ_2) (or z^1, z^2) is a minimal solution, then there are only two ways to contribute in f. First, suppose that we have an element i that is only covered by the node j ($i \in S_j^1$), although $z_j^1 = 0$, then it follows directly that $z_j^2 = 1$. If an element i is only covered by the node j ($i \in S_j^2$) although $z_j^2 = 0$, then it is possible to switch from $z_j^1 = 1$ to $z_j^2 = 1$ without increasing the objective function.

To proof that ODSCP1 is easy to solve we fill the last claim: Namely, it is easy to solve the minimum SCP for S and $\mathcal{F}^\in = \{S_i^2\}$.

I.e: W.l.o.g.(see Lemma 3.3.22 and $c = 2$) we suppose that $\Lambda_2 \subset \{i \in S : |S_i^2| = 2\}$. Therefore we obtain the optimal solution by selecting the maximum set of disjoint edges. This can be seen as an edge covering problem an therefore the problem is polynomially solvable. Consider the following graph $G(V, E)$, with the undirected edges $E = \{S_i : |S_i| = 2\}$ and $V = \bigcup_{e \in E} e \subset S$. Suppose that a minimum edge cover $E' \subset E$ for this graph is already found, then we define:

$$\Lambda_2 = \{\min\{j : S_j = e\} : e \in E'\}$$

Since $|\Lambda_2| = |E'|$ it is clear that:

$$f = |E'|$$

Ad ODSCP2: use Lemma 3.3.24. □

we will show that ODSCP2 is still NP-complete if $d \leq 2$ without the presence odd cycles. We will use a transformation of Max-2-SAT to ODSCP2.

Theorem 3.3.26. *ODSCP2 is NP-complete if $d(A_1), d(A_2) \leq 2$.*

Proof. The proof is accompanied by the following example:

$$\bigwedge_{i=1}^{5} c_i$$

3.3. Complexity Results

$$c_1 = \neg x_1 \vee \neg x_2$$
$$c_2 = \neg x_1 \vee x_2$$
$$c_3 = x_1 \vee x_2$$
$$c_4 = \neg x_3 \vee \neg x_2$$
$$c_5 = x_3 \vee x_1$$

This set of clauses is not simultaneously satisfiable. To find an assignment that realizes the maximum number of true clauses we will construct a suitable ODSCP2 instance. For the derived ODSCP2 instance we construct a sufficiently large even circuit for each variable and we claim that:

1. each circle has exactly two feasible configurations. (namely, x_i =true and x_i =false)

2. for each circle the services are "conflict free". I.e: Suppose that $C_i \subset S$ represents one of the circles. If $(S_j^1, S_j^2) = (\{v_1, v_2\}, \{v_1, v_2\}) \in \mathcal{F}$ and $v_1, v_2 \in C$ then $j \notin \Lambda_1 \cap \Lambda_2$.

3. for each clause $c_j = (l_{i_1} \vee l_{i_2})$ with $l_{i_1} \in \{x_{i_1}, \neg x_{i_1}\}$ (respectively, $l_{i_2} \in \{x_{i_2}, \neg x_{i_2}\}$) the corresponding circles use a common $f_j \in \mathcal{F}$ that has to be selected if $c_j = l_{i_1} = l_{i_2}$ =false. In other words: the clauses are represented in the connections of the circles.

We construct S and $\mathcal{F}$ in the following way:

1. $S = \bigcup_{i=1}^{n} C_i \quad$ and $\quad C_i = \{L*i+1, \ldots, L*(i+1)\}$

2. $\mathcal{F} = \bigcup_{i=1}^{n} \mathcal{F}_i \cup \bigcup_{j=1}^{m} \mathcal{F}_j$

3. $\mathcal{F}_i = \bigcup_{\lambda=1}^{\frac{L}{3}-1} (\{L*i+\lambda, L*i+(\lambda+1)\}, \{L*i+\lambda, L*i+(\lambda+1)\})$

4. For $c_j = (l_{j_1} \wedge l_{j_1})$ and x_i we define:

    - $r_1(j) = |\{c_r : r \leq j \;\wedge\; l_{r_1} = l_{j_1}\}|$
    - $r_2(j) = |\{c_r : r \leq j \;\wedge\; l_{r_2} = l_{j_2}\}|$
    - $h_1(j) = i$ if $l_{j_1} \in \{x_i, \neg x_i\}$
    - $h_2(j) = i$ if $l_{j_2} \in \{x_i, \neg x_i\}$
    - $g(l) = \begin{cases} 0, & l \in \{x_i : i \leq n\}; \\ 1, & l \in \{\neg x_i : i \leq n\}. \end{cases}$

55

3. The Double Set Cover Problem

5. $\mathcal{F}_j = \bigcup\limits_{c_j=(l_{j_1} \wedge l_{j_2})} \{(\{a(j), a(j)+1\}, \{b(j), b(j)+1\})\}$, where:
   - $a(j) = (h_1(j) - 1) * L + \frac{L}{3} + (2r_1(j) - g(l_{j_1}))$
   - $b(j) = (h_2(j) - 1) * L + \frac{2L}{3} + (2r_2(j) - (1 - g(l_{j_2})))$

6. in the last step we augment $\mathcal{F}$ by the sets that cover the missing nodes. Therefore we need to define:
   - $\mathcal{T}_1 = \bigcup\limits_{(S_j^1, S_j^2) \in \mathcal{F}} \{S_j^1\}$
   - $\mathcal{T}_2 = \bigcup\limits_{(S_j^1, S_j^2) \in \mathcal{F}} \{S_j^2\}$
   - $\mathcal{T} = \bigcup_{i=1}^{n} \bigcup\limits_{\lambda=\frac{L}{3}}^{L} \{\{L*i + \lambda, L*i + 1 + (\lambda \bmod L)\}\}$

7. $\mathcal{F} \leftarrow \mathcal{F} \cup \bigcup\limits_{T \in \mathcal{T} \setminus \mathcal{T}_1} \{(T, \phi)\} \cup \bigcup\limits_{T \in \mathcal{T} \setminus \mathcal{T}_2} \{(\phi, T)\}$

We summarize that the derived instance consist of $n' = nL$ nodes and $m' = \frac{5}{3}nL - m$ sets. To formulate a sufficient condition for L we need to define:

- $R_1^1(i) = |\{c_j : l_{j_1} = x_i\}|$ and $R_2^1(i) = |\{c_j : l_{j_1} = \neg x_i\}|$
- $R_1^2(i) = |\{c_j : l_{j_2} = x_i\}|$ and $R_2^2(i) = |\{c_j : l_{j_2} = \neg x_i\}|$

The idea is now that we choose L according to the following equations,

$$\max\{R_1^1(i), R_2^1(i), R_1^2(i), R_2^2(i)\} < \frac{L}{3} \quad \wedge \quad L \bmod 6 = 0$$

Since $f_{ODPC2} = |\Lambda_1 \cup \Lambda_2|$ counts the number of conflicts z we also suppose that:

$$z \leq \frac{L}{3} \quad \wedge \quad k = n'$$

Suppose that $z = 0$ and then we obviously found a solution for the Max-2-SAT where all clauses are satisfied. Now we have to argue that if $z > 0$ that the optimal solutions induce a perfect matching. I.e. to prove that we suppose that this is true for m clauses, and then we will observe an optimal solution for $m + 1$ clauses. If we can find a clause that causes a conflict, then we may eliminate it to get a problem with m clauses and a lower z. Since this leads us to a perfect matching we can add the conflicting clause afterwards and we come back to another optimal solution, but one that also induces a perfect matching. Therefore we can argue that we always get perfect matchings as solutions of the ODSCP2 and therefore we can directly identify the

3.3. Complexity Results

solutions as solutions for the Max-2-SAT. I.e.: If z is the smallest number for which we can get a feasible solution to this problem, we know that we need to use at least z sets (S_i^1, S_i^2) simultaneously. In Figure 3.12 the example is transformed and $L = 12$ is used. A minimal solution for $z = 1$ is given and corresponds to an optimal assignment for Max-2SAT.

□

Since the ODSCP mention the method described in Hochbaum et al. [58] that provides a 2-approximation.

We can generalize the PPMP to an optimization variant called the Minimum Conflict Perfect Matching Problem and we will apply the arguments of Theorem 3.3.26 to show that PPMP is NP-complete for low key requirements.

Definition Minimum Conflict Perfect Matching Problem (MCPMP): Given a graph $G(V, E)$ and a conflict set $W \subset E \times E$. We want to find a subset $M \subset E$ such that:

- $\min |\{\{e_i, e_j\} \in M | (e_i, e_j) \in W\}|$.

- M is a perfect matching of (V, E).

Corollary 3.3.27. *MCPMP is NP-complete even if $d(i) \leq 2$ and W is a injective partial mapping.*

Proof. In the case of Figure 3.12 we can also think of a MCPMP instance. I.e. we can think of separating the structure due to S_j^1 and S_j^2 by duplicating the nodes in S. Again, we can directly transfer the optimal solutions. Since each edge S_i^1 has only one counter part S_i^2 we can interpret W as injective partial mapping. □

We continue with the discussion of the the complexity of the the optimization versions of DSCP. Unlike the DSCP with $A_1 = A_2$ we can show that $d^- \leq 2$ doesn't lead to an easy problem.

Theorem 3.3.28. *ODSCP1 is NP-complete even if $A_1 = A_2$ and $d^+ \leq 3$ and $d^- \leq 2$.*

Proof. We polynomially reduce VERTEX COVER to ODSCP1: We will construct an instance with $d^- = 2$ and we can interpret each $i \in S$ as an edge $S_{i_1}, S_{i_2} \in \mathcal{F}$. Now, the idea is to directly transfer the VERTEX COVER instances. Since Garey et al. [45] show that VERTEX COVER remains NP-complete if the maximum degree is 3 we can sustain $d^+ \leq 3$ and $d^- \leq 2$ for the instance. If we can also transfer k since the error doesn't matter. I.e: we will show that the vertex cover problem is feasible if and only if ODSCP1 is feasible. A solution to the vertex cover $C \subset V$ can be transferred to a solution $\Lambda_1 = S$ and $\Lambda_2 = \phi$. On the other hand, if (Λ_1, Λ_1) is a solution to the ODSCP1 then we know that $\Lambda_1 \cup \Lambda_1$ is a solution to the vertex cover problem. □

57

3. The Double Set Cover Problem

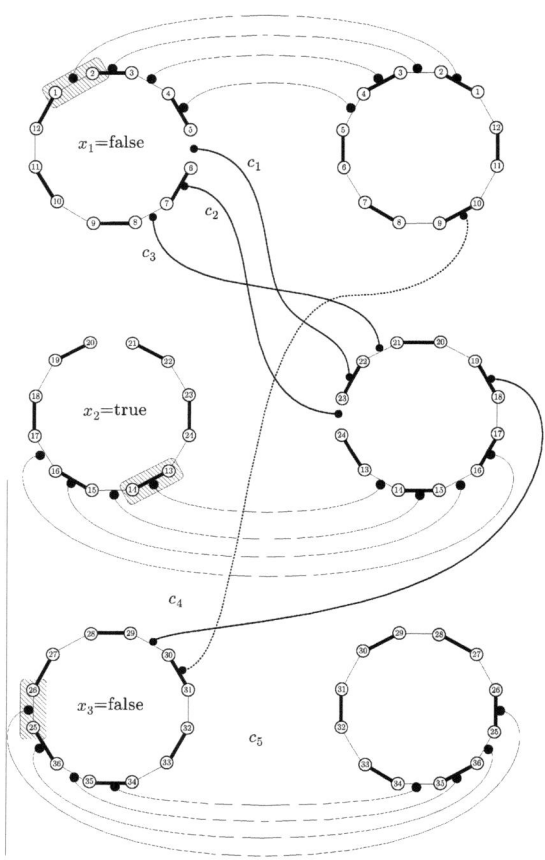

Figure 3.12.: Example of a transformation from Max-2-SAT to ODSCP2

Also for the ODSCP2 we can conclude that:

Corollary 3.3.29. *ODSCP2 is NP-complete even if $A_1 = A_2$ and $d^+ \leq 3$ and $d^- \leq 2$.*

Proof. reduction of vertex cover to ODSCP1: we follow the arguments in the proof of Theorem 3.3.28 and set $k' = 2*k$. □

Like Figure 3.11 that summarizes the results for the DSCP we also prepare a Figure 3.13 that gives an overview of the complexity results for the optimization versions of DSCP. Here we classify NP-hard and polynomially solvable sub problems.

3.4. Inference of Gene Regulatory Networks

In this section we will present the joint work with G. Lulli [82] that deals with the inference of *Gene Regulatory Network Problem*. This problem originally motivated the discussion about the *Double Set Cover Problems*. The aim of this section is to provide a mathematical tool that filters out information from a large amount of putative regulations.

3.4.1. Introduction

Recent technological advances enable biomedical investigators to observe the genome of entire organisms in action by simultaneously measuring the level of activation of thousands of genes under the same experimental conditions. This technology, known as microarrays, provides unparalleled discovery opportunities and is reshaping biomedical sciences. One of the main aspects of this revolution is the introduction of computationally intensive data analysis methods in biomedical research. These methods have already contributed to the discovery of a large number of genes and their regulatory sites. Much less is known, however, about the functioning of the regulatory systems of which the individual gene and its interactions form a part, e.g., see [8, 93]. With few exceptions, all the cells in an organism contain the same genetic material. To better understand how genes are implicated in the control of intracellular and intercellular processes, it is of fundamental importance to gain insight into gene function and operation, and in their functional linkages [2].

Gene regulatory networks regulate the expression of thousands of genes. Uncovering such networks is essential for understanding how genomic expression programs unfold during developmental processes, how the molecular machinery of cells works to respond adequately to environmental clues and to maintain homeostasis, and, consequently, how to manipulate these processes to human advantage. Hence, gaining an understanding of the emergence of complex

3. The Double Set Cover Problem

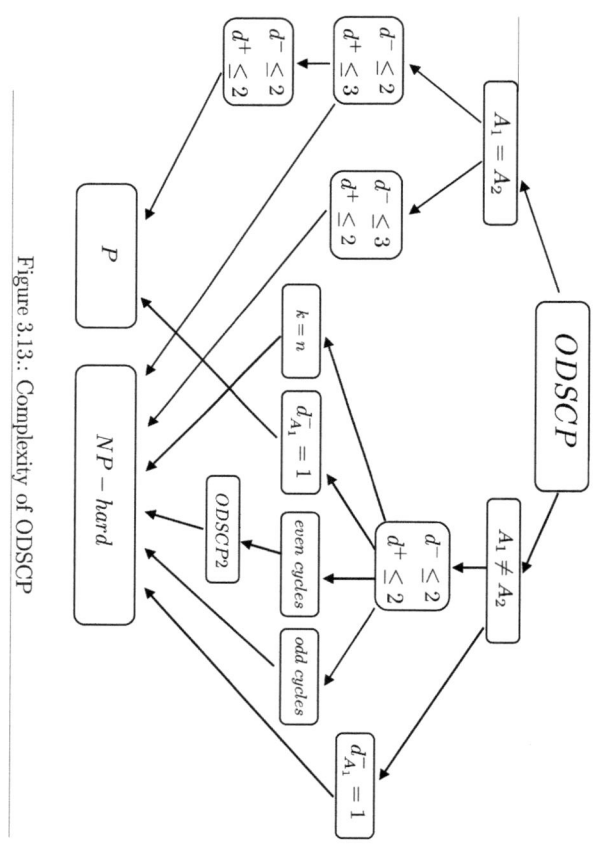

Figure 3.13.: Complexity of ODSCP

3.4. Inference of Gene Regulatory Networks

patterns of behavior from the interactions between genes in a regulatory network poses a huge scientific challenge with potentially high industrial pay-offs.

Several methods have been proposed to reconstruct gene regulatory networks from gene expression data. The goal of these methods is to produce a high-fidelity representation of the cellular network topology as a graph, whose nodes represent genes and arcs represent direct regulatory interactions [85]. There is a wide spectrum of techniques and criteria to define an arc. de Jong [23], Filkov [39], Friedman [43], Lee [78], van Someren [122] among others, surveyed most of the existing methods to date. Bayesian network [14, 42, 95], Boolean networks [67, 115], systems of Ordinary Differential Equations [11, 22, 48, 103, 123] and Computational Algebra methodologies [77] define arcs as parent-child relationships between mRNA abundance levels that are most likely to explain the data. Integrative methods [1, 10] use independent experimental clues to define arcs as those showing evidence of physical interactions. Statistical methods [38, 94, 110, 120] identify arcs with the strongest statistical associations between mRNA abundance levels. All methods show a trade-off between the level of approximation of the real network and the level of computational difficulties. Some of them provide models that might be considered as a coarse-grained approximation to the "real" network, but they can be applied to large scale networks, others focus on subnetworks, e.g. [3, 127]. Although all these approaches may seem disjointed, they might not be as far apart as it appears and could serve complementary roles.

Most of the cited methods face two main challenges. The first one involves statistical robustness. Building a network that involves thousands of genes from several dozen examples of their joint expression levels is extremely problematic. These examples are not enough to distinguish between true correlations and spurious ones. The second and more difficult challenge is the biological interpretability of the results and some questions arise on how to distinguish regulation from co-expression and direct regulations from indirect ones. Whereas the ultimate goal is to identify the direct regulation of targets by transcription factors, experience shows that the mentioned methods also reveal many other indirect relations. Therefore, we often come up with a putative network containing putative relationship between genes.

In the paper [82], we focussed on the problem of pruning a putative regulatory network obtained from gene expression data sets by applying appropriate inference methods. We formalize the problem of identifying a smaller set of interesting candidate regulatory elements by means of a mathematical program. A similar problem has been tackled by Chen et al. [12], who proposed an approximation algorithm. As well as Chen et al. [12], we do not assert that we identify the regulatory network as result of our computation, but, we believe that our approach quickly enables biologists to identify and visualize interesting features from raw expression array data sets. The proposed method is also intended to generate a "refined draft" of large gene

3. The Double Set Cover Problem

regulatory networks, on which further, more local, analysis can be based. It can also be used as "warm start" for several of the approaches listed above which assume that an initial network is given, e.g., Kholodenko et al. [68] and Kim et al. [69] to mention just a few. Moreover, even if an initial guess on the network topology is not required, in many cases, a refined draft network can be useful to guide procedures designed to infer gene regulatory network, e.g. [24].

As a real world test to validate our approach, we used the 17-point time series data set measuring the expression level of each of 6601 different genes from the *Saccharomyces cerevisiae* genome, first published in [15] (available at http://genome-www.stanford.edu/ /cellcycle/data/rawdata/).

The rest of the section is organized as follows. In §3.4.2 we present the mathematical programming formulation of the model, which is shown, in §3.4.4, to be NP-complete by reduction of the SET COVER decision problem. In §3.4.3, we give an equivalent Mixed Integer Programming (MIP) formulation. To solve large instances, in §3.4.5 we propose a metaheuristic algorithm based on the Ant Colony Optimization (ACO) concept. The results of computational analysis on instances of the GRNP problem are reported in §3.4.6. Finally, §3.4.9 contains the conclusions.

3.4.2. The Gene Regulatory Network Problem

Gene regulatory networks represent the activation/inhibition influence of gene products upon the expression of other genes. Every gene has one or more activators (resp. inhibitors), i.e., biochemical signals which are necessary to start (resp. to prevent) the transcription of the gene. Any reverse engineering algorithm for inferring gene regulatory networks might compute network topologies containing spurious correlations and indirect relations due to the reduced amount of data and measurement noise.

The mathematical model we present, is purposely designed to generate networks in which a relatively small number of regulators explain the expression of all other genes. These networks are obtained by pruning putative gene relations, which are derived from gene expression raw data sets by applying appropriate inference methods. Putative gene regulatory networks usually contain genes which activate the expression of some genes and inhibit the expression of others. Loosely speaking, we say that these genes are both activators and inhibitors. Even though the presence of such genes is consistent with biological evidence, e.g., [47, 96], few examples are known and their number should be rather small. In our model, we *implicitly* minimize the number of these genes by deleting either all the activation or all the inhibition influences of the gene upon the others. However, not all these influences can be deleted in order to promote or to repress the expression of all the genes of the network. Furthermore, we call neutral those

3.4. Inference of Gene Regulatory Networks

genes, though playing a role in biochemical intracellular and intercellular processes, do not have any activation/inhibition influence upon other genes of the network.

In mathematical terms, we represent the putative gene network by a graph $\mathcal{G}(N, E', E'')$. N, the set of nodes, represents the gene products, $E' \subset N \times N$ and $E'' \subset N \times N$, the sets of arcs, represent the sets of putative activation and inhibition influences respectively. For clarity of exposition, we call E' the set of activating arcs and E'' the set of inhibiting arcs. If a gene i activates (inhibits) a gene j then there exists and arc $[i, j] \in E'(E'')$. Gene i is a predecessor of gene j while gene j is a successor of gene i. The decision problem is to label nodes either as activators, inhibitors or neutral which explains the expression of all the genes while minimizing the number of "irregular" influences. An irregular influence is an arc whose label differs from the label of the parent node, i.e., an activation arc $\in E'$ (resp. inhibition arc $\in E''$) whose parent node is labeled inhibitor (resp. activator). The presence of an irregular arc means that a gene is both activator and inhibitor.

Our problem is similar to the maximum gene regulation problem proposed by Chen et al. [12]. However, with respect to their model, we also consider neutral gene products, whose number it is supposed to be bounded below by a known parameter M. Moreover, in our model gene products can be both activators and inhibitors, i.e., they can induce both activation and inhibition influences, even though we implicitly reduce the number of such genes.

We assign to each node (gene) of the network, two binary variables: $z_i^{(A)}$ and $z_i^{(I)}$. The former labels node i as activator, the latter as inhibitor. We also use a binary decision variable for each arc of the putative network to discern if the arc is either held or removed from the graph. For the convenience and clarity for the reader, we here summarize the decision variables:

$$z_i^{(A)} = \begin{cases} 1 & \text{if node } i \text{ is labeled as activator,} \\ 0 & \text{otherwise.} \end{cases}$$

$$z_i^{(I)} = \begin{cases} 1 & \text{if node } i \text{ is labeled as inhibitor,} \\ 0 & \text{otherwise.} \end{cases}$$

$$x_{ij} = \begin{cases} 1 & \text{if arc } [i,j] \in E' \text{ is held in the network,} \\ 0 & \text{otherwise.} \end{cases}$$

$$y_{ij} = \begin{cases} 1 & \text{if arc } [i,j] \in E'' \text{ is held in the network,} \\ 0 & \text{otherwise.} \end{cases}$$

As described above, we minimize the number of irregular arcs.

$$\min \sum_{i \in N} (1 - z_i^{(I)}) \sum_{j:[i,j] \in E''} y_{ij} + \sum_{i \in N} (1 - z_i^{(A)}) \sum_{j:[i,j] \in E'} x_{ij}$$

3. The Double Set Cover Problem

$$
\begin{align}
s.t. \quad & \sum_{i:[i,j] \in E'} x_{ij} \geq 1 & \forall j \in N & \quad (3.31) \\
& \sum_{i:[i,j] \in E''} y_{ij} \geq 1 & \forall j \in N & \quad (3.32) \\
& z_i^{(A)} + z_i^{(I)} \leq 1 & \forall i \in N & \quad (3.33) \\
& \sum_{i \in N} z_i^{(A)} + z_i^{(I)} \leq |N| - M & & \quad (3.34) \\
& \sum_{[i,j] \in E'} x_{ij} + \sum_{[i,j] \in E''} y_{ij} \leq d_G^+(i) \cdot (z_i^{(A)} + z_i^{(I)}) & \forall i \in N & \quad (3.35) \\
& x_{ij} \in \{0,1\} & \forall [i,j] \in E' & \\
& y_{ij} \in \{0,1\} & \forall [i,j] \in E'' & \\
& z_i^{(A)}, z_i^{(I)} \in \{0,1\} & \forall i \in N &
\end{align}
$$

Sets of constraints (3.31) and (3.32) force each gene to have at least one activator and one inhibitor. That is, each node has at least one inhibiting arc and one activating arc among all the incoming arcs in the final network. Constraints (3.33) impose that each gene can be labeled either as activator, inhibitor or neutral. A gene is neutral if both the z variables take value 0. The number of activator and inhibitor genes is bounded above by constraints (3.34). More precisely, constraints (3.34) impose that at least M genes products are neutral. In constraints (3.35), $d_G^+(i)$ denotes the outdegree of node i. These constraints impose that neutral genes do not have influences upon other genes. In this case, the right hand side of constraints (3.35) is equal to 0. These constraints are redundant for nodes labeled as either activators or inhibitors.

Note that, the sets of constraints (3.31) and (3.32) are satisfied if each node of the network has at least one incoming arc labeled as activator and one as inhibitor, i.e., if the following condition holds:

$$\{j \mid \exists \, [i_1, j] \in E' \wedge [i_2, j] \in E''\} = N$$

In case this condition does not hold, it is purposely enforced by adding some dummy activation/inhibition arcs, see §3.4.8 for details. Note also that these constraints, i.e., constraints (3.31) and (3.32), are coherent with biological knowledge especially if we consider both transcriptional and translational levels in molecular cell cycle mechanisms [72, 106].

3.4.3. A Mixed-Integer Linear Programming Formulation

The formulation presented in §3.4.2 is bilinear ([BIL]). However, the objective function can be linearized using one of the approaches proposed for this class of problem, e.g. see Sherali and Adams [111]. Herein, we present a more compact Mixed-Integer Linear Program (MIP)

3.4. Inference of Gene Regulatory Networks

formulation for the gene regulatory problem, named [GRNP-MIP], equivalent to [BIL].

The [GRNP-MIP] formulation appends the same decision variables, $z_i^{(A)}$ and $z_i^{(I)}$, to each node of the network (gene). Again, the former decision variable labels node i as activator, the latter as inhibitor. Additional decision variables are used to take into account "node regulation". We say that a gene is *regularly activated* if it is activated by a gene labeled as activator. In graph theoretical terms, it means that at least one of the adjacent (parent) nodes by means of activating arcs is labeled activator. The case of a *regularly inhibited* gene is similar.

In summary, the node regulation decision variables are formally defined:

$$\delta_j^{(A)} = \begin{cases} 1 & \text{if node } j \text{ is regularly activated,} \\ 0 & \text{otherwise} \end{cases}$$

$$\delta_j^{(I)} = \begin{cases} 1 & \text{if node } j \text{ is regularly inhibited,} \\ 0 & \text{otherwise} \end{cases}$$

The objective function minimizes the number of nodes irregularly activated/inihbited, which is of course equivalent to maximizing the number of regular influences.

$$\min_{z,\delta} \underbrace{\sum_{i \in N} (1 - \delta_i^{(A)}) + (1 - \delta_i^{(I)})}_{f_{GRNP-MIP}(z,\delta):=}$$

$$\begin{align}
\text{s.t.} \quad & \sum_{i:[i,j]\in E'} z_i^{(A)} \geq \delta_j^{(A)} & \forall j \in N & \quad (3.36) \\
& \sum_{i:[i,j]\in E''} z_i^{(I)} \geq \delta_j^{(I)} & \forall j \in N & \quad (3.37) \\
& z_i^{(A)} + z_i^{(I)} \leq 1 & \forall i \in N & \quad (3.38) \\
& \sum_{i \in N} z_i^{(A)} + z_i^{(I)} \leq |N| - M & & \quad (3.39) \\
& \sum_{i:[ij]\in E'} (z_i^{(I)} + z_i^{(A)}) \geq 1 & \forall j \in N & \quad (3.40) \\
& \sum_{i:[ij]\in E''} (z_i^{(I)} + z_i^{(A)}) \geq 1 & \forall j \in N & \quad (3.41) \\
& z_i^{(A)}, z_i^{(I)} \in \{0,1\} & \forall i \in N & \quad (3.42) \\
& \delta_i^{(A)}, \delta_i^{(I)} \in \{0,1\} & \forall i \in N & \quad (3.43)
\end{align}$$

The sets of constraints (3.36) and (3.37) fix the node regulation decision variables, δ. If none of the parent nodes by means of activating (inhibiting) arcs is labeled activator (inhibitor) then $\delta_j^{(A)}$ ($\delta_j^{(I)}$) is forced to take value 0. Constraints (3.38) and (3.39) play the same role as constraints (3.33) and (3.34) in [BIL]. Constraints (3.40) and (3.41) impose that each gene has at least one activator and one inhibitor.

3. The Double Set Cover Problem

Note that, the left hand side of constraints (3.36) and (3.37) is always a non-negative integer value. In view of the objective function, the decision variables δ will always attain their maximum value in correspondence of any optimal solution; therefore they will always be either 0 or 1. This justifies the following claim:

Claim 1: *The optimal solution of the relaxed formulation of [GRNP-MIP], obtained relaxing the integer requirements on δ decision variables, is integer, i.e., feasible and thus optimal for [GRNP-MIP].*

Note also that, the optimal solution of [GRNP-MIP] is univocally identified by the z variables.

Lemma 3.4.1. *Given a feasible solution (z, δ) of [GRNP-MIP] there exists a feasible solution (z, x, y) of [BIL] such that $f_{BIL}(z, x, y) \leq f_{GRNP-MIP}(z, \delta)$.*

Proof. Consider a feasible solution of [GRNP-MIP], (z, δ). A feasible solution of [BIL] (z, x, y) is constructed as follows.

We let the z variables of the [BIL] solution to take the same values of those of the (z, δ) solution. Hence, constraints (3.33) and (3.34) hold since they are the same constraints as (3.38) and (3.39) and involve only the z variables. As far as the x and y variables, we set to 1 all the x (y) variables referring to regular activations (inhibitions), i.e. activating (inhibiting) arcs whose parent node is labeled activator (inhibitor). Thus,

$$\forall i \in N : z_i^{(A)} = 1 \implies x_{ij} = 1 \quad \forall [i,j] \in E'$$
$$\forall i \in N : z_i^{(I)} = 1 \implies y_{ij} = 1 \quad \forall [i,j] \in E''$$

All these edges do not increase the objective function $f_{BIL}(z, x, y)$ since they represent regular activations and inhibitions. Moreover, nodes regulated by these arcs are regularly activated and/or inhibited and thus satisfy conditions (3.31) and/or (3.32).

Now, we consider not regularly regulated nodes, i.e., nodes for which either $\sum_{i:[i,j]\in E'} z_i^{(A)} = 0$, $\sum_{i:[i,j]\in E''} z_i^{(I)} = 0$, or both.

Suppose that there exists a node j such that $\sum_{i:[i,j]\in E'} z_i^{(A)} = 0$ which implies $\delta_j^{(A)} = 0$ (by constraint (3.36)). By constraints (3.40), which hold by assumption, it follows that $\sum_{i:[i,j]\in E'} z_i^{(I)} \geq 1$ or equivalently $\{i : [i,j] \in E' \wedge z_i^{(I)} = 1\} \neq \phi$. Hence, one of the irregular activators has to be chosen to activate node j, for instance the putative regulation with the smallest index $(i_{min}^{(A)})$. The corresponding x variable is set to 1 ($x_{i_{min}^{(A)}j} = 1$), while all the others are set to zero. So we have:

3.4. Inference of Gene Regulatory Networks

$$x_{ij} = \begin{cases} 1 & \text{if } i = i_{min}^{(A)} = \min\{i : [i,j] \in E' \wedge z_i^{(I)} = 1\} \\ 0 & \text{otherwise.} \end{cases}$$

A node j, which is not regularly inhibited (i.e., $\sum_{i:[i,j]\in E''} z_i^{(I)} = 0$), is handled in a similar manner by fixing the y variables as follows:

$$y_{ij} = \begin{cases} 1 & \text{if } i = i_{min}^{(I)} = \min\{i : [i,j] \in E'' \wedge z_i^{(A)} = 1\} \\ 0 & \text{otherwise.} \end{cases}$$

Both the x and y variables of the solution so constructed, satisfy constraints (3.31), (3.32) and (3.35); therefore the solution (z, x, y) is feasible. The inequality

$$f_{BIL}(z,x,y) \leq f_{GRNP-MIP}(z,\delta)$$

immediately follows by noting that for every node of $[GRNP\text{-}MIP]$, not regularly regulated (δ-variable $= 0$), we select exactly one irregular arc for $[BIL]$. □

Lemma 3.4.2. *Every feasible solution of [BIL] satisfies the sets of constraints (3.40) and (3.41) of [GRNP-MIP].*

Proof. Let (z,x,y) be a feasible solution of $[BIL]$ which violates constraints (3.40), i.e., $\sum_{i:[i,j]\in E'}(z_i^{(I)} + z_i^{(A)}) = 0$ for at least one node j. Since constraints (3.31) hold, then there exists at least one node $\tilde{i}$ that activates node j, $(x_{\tilde{i}j} = 1)$. But $x_{\tilde{i}j}$ also occurs on the left hand side of (3.35), therefore either $z_{\tilde{i}}^{(I)} = 1$ or $z_{\tilde{i}}^{(A)} = 1$, which contradicts our assumption. Similar argument can be used to show that constraints (3.41) are also satisfied. □

Lemma 3.4.3. *Given a feasible solution (z,x,y) of [BIL], there exists a feasible solution of [GRNP-MIP], (z,δ), such that $f_{GRNP-MIP}(z,\delta) \leq f_{BIL}(z,x,y)$.*

Proof. Suppose (z,x,y) is a feasible solution of $[BIL]$. We let to the z decision variables of the $[GRNP\text{-}MIP]$ solution to take the same values of those of the (z,x,y) solution. These z variables trivially satisfy constraints (3.38), (3.39) and they also satisfy constraints (3.40) and (3.41) by Lemma 3.4.2. Therefore, to come to a feasible solution of $[GRNP\text{-}MIP]$ we only need to construct appropriate δ decision variables. We fix the δ variables as follows:

$$\delta_j^{(A)} = \begin{cases} 1 & \text{if } \sum_{i:[i,j]\in E'} z_i^{(A)} \geq 1 \\ 0 & \text{otherwise.} \end{cases}$$

$$\delta_j^{(I)} = \begin{cases} 1 & \text{if } \sum_{i:[i,j]\in E''} z_i^{(I)} \geq 1 \\ 0 & \text{otherwise.} \end{cases}$$

3. The Double Set Cover Problem

Hence, the (z, δ) solution is feasible for [GRNP-MIP] by construction.

We now verify that the inequality $f_{GRNP-MIP}(z, \delta) \leq f_{BIL}(z, x, y)$ holds. The δ variables are responsible for increasing the objective function $f_{GRNP-MIP}(z, \delta)$. Suppose that $\delta_j^{(A)} = 0$ (i.e., $\sum_{i:[i,j] \in E'} z_i^{(A)} = 0$). The $\delta_j^{(I)} = 0$ (i.e., $\sum_{i:[i,j] \in E''} z_i^{(I)} = 0$) case is analogous. $\delta_j^{(A)} = 0$ increases $f_{GRNP-MIP}(z, \delta)$ by one unit. By virtue of constraints (3.31) and (3.35), at least one node i, labeled inhibitor ($z_i^{(I)} = 1$), activates node j ($x_{ij} = 1$), i.e., we have at least one irregular influence which augments $f_{BIL}(z, x, y)$ by one unit. From this, the inequality immediately follows. □

From the lemmas listed above it is straightforward to prove by contradiction the following theorem.

Theorem 3.4.4. *At the optimum both [GRNP-MIP] and [BIL] show the same value, i.e., $f^*_{GRNP-MIP} = f^*_{BIL}$. Moreover, given an optimal solution of one formulation is possible to construct an optimal solution of the other.*

3.4.4. Complexity of the GRNP

To prove the NP-completeness of GRNP, we show that any instance of the SET COVER decision problem, which is is NP complete [46], polinomially transforms to an instance of GRNP (see (3.36)-(3.43)).

Definition SET COVER Decision Problem. Given a finite set $S = \{1 \ldots m\}$ and a collection of subsets of S, $\mathcal{C} = \{S_i \subset S | i = 1 \ldots n\ \}$, does $\mathcal{C}$ contain a cover of S of size k ?

Or equivalently, does a subset $\Lambda \subset \{1 \ldots n\}$ exist such that $S = \bigcup_{i \in \Lambda} S_i$?

We say that an instance of the SET COVER Decision Problem is not trivial if the following conditions hold:

$$S = \bigcup_{i=1}^{n} S_i \qquad |S| \geq 2.$$

Definition GRNP Decision Problem. Given an instance of the GRNP problem, i.e., a graph $G(N, E', E'')$, does a labeling of nodes exist such that the GRNP has at most $k+1$ irregular influences?

Theorem 3.4.5. *GRNP is NP-complete even if the number of neutral nodes is set to 0 and the outdegree of either activation or inhibition arcs is 1 for all the nodes.*

Proof. We transform instances of the SET COVER decision problem into restricted instances of the GRNP decision problem. More specifically, we consider the following two restrictions:

3.4. Inference of Gene Regulatory Networks

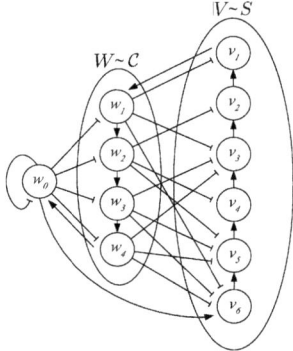

Figure 3.14.: The derived graph for the following SET COVER instance: $S = \{v_1, v_2, v_3, v_4, v_5, v_6\}$ and $\mathcal{C} = \{S_1, S_2, S_3, S_4\}$ where $S_1 = \{v_1, v_3, v_6\}, S_2 = \{v_2, v_4, v_5\}, S_3 = \{v_3, v_4, v_6\}$ and $S_4 = \{v_3, v_4, v_6\}$.

i the outdegree of activating arcs is equal to 1 for all the nodes;

ii the minimum number of neutral nodes is set to 0.

In view of restriction *i*, any feasible instance of the GRNP problem has all the nodes with both outdegree and indegree of activating arcs equal to 1. Moreover, as a consequence of both the restrictions, all the nodes are labeled either as activators or inhibitors in any feasible solution of the GRNP problem.

Given an arbitrary, non trivial, instance of the SET COVER decision problem, we derive a restricted instance of GRNP by considering the following graph $G(N, E', E'')$. $N = V \cup W \cup \{w_0\}$ is the set of nodes, where $V = \{v_1, \ldots, v_m\}$ is a copy of S and $W = \{w_1, \ldots, w_n\}$ is a copy of $\mathcal{C}$. E' and E'' are the sets of arcs. E' is the set of putative activation arcs and is restricted to an arbitrary assignment of the nodes (e.g. by a hamiltonian cycle). $E'' := E_1'' \cup E_2''$ is the set of putative inhibition arcs and is defined as follows:

- $E_1'' = \bigcup_{i=1}^{n} \{[w_i, v_j] | j \in S_i\}$.

- $E_2'' = \{[w_0, w_i] | i = 1, \ldots, n\}$.

Figure 3.14 shows the transformation of an instance of the SET COVER problem to an instance of [GRNP]. The arcs with a T-ending represent putative inhibiting influences and the arrows represent putative activating influences. The outgoing arcs from node w_i represent the elements of the set S_i. For instance, the arc $[w_2, v_4] \in E''$ means that $v_4 \in S_2$. We can also see that w_0

putatively inhibits itself and all the elements of the set W. For the set E' (the arrows) we have chosen an arbitrary hamiltonian cycle.

Before going on to prove the Theorem, we first observe the following facts:

Claim 2. *In any optimal solution of the constructed GRNP instance, all the nodes in S are labeled activators.*

Note that, all the nodes in S exert only activation influences. If one of these nodes is labeled as inhibitor then it is possible to improve the solution, i.e., to reduce the number of irregular influences, by simply switching the label of the node.

Claim 3. *In any optimal solution of the constructed GRNP instance, we may assume that the irregular influences are activating influences.*

By restriction, the outdegree of activating influences is one for all the nodes. Therefore, if a node exerts at least one inhibition influence then it is not inconvenient to label the node as inhibitor. In case the node does not inhibit any other node then, by labeling the node as activator, it does not induce any irregular influence.

Here, we show that a "YES" answer to the GRNP decision problem exists if and only if there is a "YES" answer to the SET COVER decision problem, i.e., there exists a solution Λ for the SET COVER, with $|\Lambda| = k$.

($\Rightarrow$) Suppose that there exists a "YES" answer to the GRNP decision problem. Without loss of generality, suppose that the $k+1$ irregular influences are activating influences, see Claim 3. These influences are promoted by nodes labeled inhibitors, more precisely node w_0 plus k nodes of the set W, see Claim 2. Therefore all the inhibiting influences are regular which implies that there exists a solution for the SET COVER with cardinality k.

($\Leftarrow$) Suppose the answer to the SET COVER decision problem is "YES". We label both the k nodes of W which cover the set S and node w_0 as inhibitors. Thus, all the nodes of the GRNP instance are regularly inhibited. All the other nodes are labeled as activators thus resulting in no irregular activations. Hence, all the irregular influences, activations, are $k+1$ and are those induced by nodes labeled as inhibitors. From this, we have constructed a "YES" instance of the GRNP decision problem. □

3.4.5. Using ACO for solving the GRNP

To solve large instances of the GRNP problem, we implemented a heuristic procedure that follows the Ant Colony Optimization (ACO) idea. ACO is a relatively new bio-inspired population based metaheuristic, and it has already been effectively applied to various combinatorial

3.4. Inference of Gene Regulatory Networks

optimization problems [30]. At each iteration a set (colony) of agents (ants) individually constructs solutions. The construction of each solution is based on both heuristic information (visibility) and adaptive memory (pheromone). The pheromone gives a learning effect from past iterations.

The idea of ACO came from behavioral studies on ants, which investigated the foraging habits of ants [54]. Experiments on real ants showed how pheromone trails are used to find the shortest path between the nest and the food source.

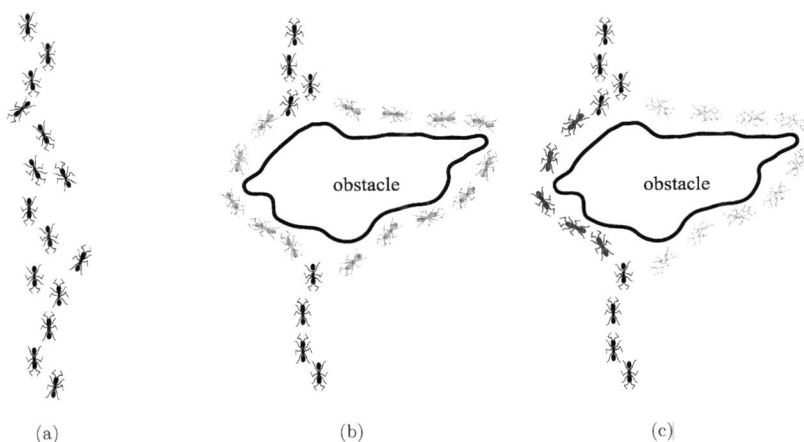

Figure 3.15.: Ants when choosing the shortest way between their nest and a food source

Figure 3.15 displays three situations: In the first one, the ants have a direct connection to the food source (a). An obstacle is placed between the nest and the food source, to offer two new routes to the food source - one longer and one shorter. When encountering the new situation the ants use both the routes with same probability to avoid the obstacle (b). In the same time interval the shorter route is used more often than the longer one. Therefore, the pheromone level of the shorter route increases more rapidly and almost all the ants are going to use it (c), see [28, 29].

Algorithm 3 reports the pseudo code of the method applied to the GRNP problem. The way how procedures are implemented and parameters are set reflects the customization of ACO to the GRNP problem. The algorithm consists of two main nested loops.

In the inner loop, the procedure $ant(\eta, \tau)$ constructs a solution by iteratively adding a new regulating element to the partial solution. A partial solution is represented by both the set of

3. The Double Set Cover Problem

Algorithm 3 ACO

$v_{min} = 2 \cdot |N|$
$\tau_i^{(A)} = \frac{|N|-M}{2}$
$\tau_i^{(I)} = \frac{|N|-M}{2}$
for $(1 \dots n\_iterations)$ **do**
   $\widetilde{v_{min}} = 2 \cdot |N|$
   for $(1 \dots n\_ants)$ **do**
      $s = (\tilde{I}, \tilde{A}, v) = ant(\eta, \tau)$
      $evaporate(s, \rho_1)$
      $s = local\_search(s)$
      if $v < v_{min}$ **then**
         $s_{min} = s$
      end if
      if $v < \widetilde{v_{min}}$ **then**
         $\widetilde{s_{min}} = s$
      end if
   end for
   $pheromone\_update(\widetilde{s_{min}}, s_{min})$
   $evaporate(\rho)$
end for
return s_{min}

---

selected activators $\tilde{A} \subset N$ and the set of selected inhibitors $\tilde{I} \subset N$. A new regulator element i ($\tilde{A} \leftarrow \tilde{A} \cup \{i\}$ or $\tilde{I} \leftarrow \tilde{I} \cup \{i\}$) is selected from the set of not yet labeled nodes $\tilde{R}$ ($= N \setminus \tilde{I} \cup \tilde{A}$) by applying either a greedy rule, with a 70% probability, or a random decision rule (roulette wheel), with a probability of 30%.

- The greedy rule selects the element with the largest value of pheromone (τ) plus visibility (η). In formula:

$$\tilde{X} \leftarrow \tilde{X} \cup \{i\} \Leftrightarrow (i, X) = \underset{(j,Y) \in \tilde{R} \times \{I, A\}}{\arg\max} \{\eta_j^{(Y)}(\tilde{Y}) + \tau_j^{(Y)}\}. \quad (X \in \{I, A\})$$

In our specific implementation of the ACO algorithm, the visibility of element i is its covering number, that is, the number of additional elements regulated by element i. In other words, those elements which are already regulated by genes(elements) included in the current partial solution are not counted for the covering number of element i.

$$\eta_i^{(A)}(\tilde{A}) = |\{j \in N : [i,j] \in E'\} \setminus \{j \in N : [k,j] \in E', \ k \in \tilde{A}\}|$$
$$\eta_i^{(I)}(\tilde{I}) = |\{j \in N : [i,j] \in E''\} \setminus \{j \in N : [k,j] \in E'', \ k \in \tilde{I}\}|$$

3.4. Inference of Gene Regulatory Networks

- The roulette wheel rule selects element i, $\tilde{I} \leftarrow \tilde{I} \cup \{i\}$ (resp. $\tilde{A} \leftarrow \tilde{A} \cup \{i\}$) with a probability proportional to $\eta_i^{(I)}(\tilde{I}) + \tau_i^{(I)}$ (resp. $\eta_i^{(A)}(\tilde{A}) + \tau_i^{(A)}$).

$$\mathbf{P}(\tilde{X} \leftarrow \tilde{X} \cup \{i\}) = \frac{\eta_i^{(X)}(\tilde{X}) + \tau_i^{(X)}}{\sum_{(j,Y) \in \tilde{R} \times \{I,A\}} \eta_j^{(Y)}(\tilde{Y}) + \tau_j^{(Y)}} \qquad (X \in \{I, A\})$$

Once a new solution s is constructed, the pheromone of each element in sets $\tilde{A}$ and $\tilde{I}$ is slightly reduced by a constant multiplicative factor $0 < \rho_1 < 1$ (near 1), thus lessening the probability for solution s to be constructed twice (procedure $evaporate(s, \rho_1)$). Then, the $local\_search(s)$ procedure is executed to improve the quality of the solution. $local\_search(s)$ implements a first improvement 2-exchange local search procedure which first recovers from possible infeasibilities. The inner loop is repeated a number of times equal to the number of ants.

Before to start a new iteration - outer loop of the algorithm - the pheromone is updated in two steps.

- In the first step, procedure $pheromone\_update(\widetilde{s_{min}}, s_{min})$ is invoked to update only the pheromone of nodes selected (labeled) in the solution. $\widetilde{s_{min}} = (\tilde{A}, \tilde{I}, \tilde{v})$ denotes the best solution computed during the last iteration and $s_{min} = (A_{min}, I_{min}, v_{min})$ denotes the best solution computed so far in the execution of the algorithm. To update the pheromone, by means of the following formulas (3.44) and (3.45), it is used either $\widetilde{s_{min}}$, if its quality is considerably good, or s_{min}.

$$\forall i \in \tilde{A}: \quad \tau_i^{(A)} \leftarrow \tau_i^{(A)} + \xi \cdot (N - \tau_i^{(A)}) \qquad (3.44)$$
$$\forall i \in \tilde{I}: \quad \tau_i^{(I)} \leftarrow \tau_i^{(I)} + \xi \cdot (N - \tau_i^{(I)}) \qquad (3.45)$$

- In the second step - procedure $evaporate(\rho)$ -, the pheromone of all the nodes is reduced by a constant multiplicative factor ρ. ρ is set to a value such that the net change of pheromone computed over all the nodes is null.

To tune the algorithm parameters, we carried out computational tests on small size instances ($|N| = 50$). The size of the colony, i.e., number of ants (n_ants), is set to 50. High initial values of the pheromone ($\tau_0 \geq \frac{|N|-M}{2}$) give better performances, since they guarantee a longer exploration phase. For the local pheromone evaporation we consider $\rho_1 = 0.99$. In the $pheromone\_update(\widetilde{s_{min}}, s_{min})$ procedure, both the value of ξ and the solution used to update

3. The Double Set Cover Problem

Table 3.1.: Settings of the ACO parameters

Case	SU	ξ
$\widetilde{v} \leq v_{min}$	$\widetilde{s_{min}}$	0.5
$v_{min} < \widetilde{v} \leq 1.1 \cdot v_{min}$	$\widetilde{s_{min}}$	0.05
$\widetilde{v} > 1.1 \cdot v_{min}$	s_{min}	0.05

SU, solution used to update the pheromone.

the pheromone depend on the quality of $\widetilde{s_{min}}$ with respect to s_{min}. In Table 3.1, we report the possible cases where we denoted with $\widetilde{v}$ and v_{min} the objective function value of $\widetilde{s_{min}}$ and s_{min} respectively.

3.4.6. Computational Experience

In this section, the computational results of the proposed method are reported. In Figure 3.16, we schematically depict the complete procedure to generate a reduced and coherent regulatory network from raw expression array data sets. In grey, we highlight the parts of the procedure addressed in this section. A putative gene regulatory network is obtained from microarray expression data by applying an appropriate inference method, see §3.4.8 for details. Given the putative regulatory network we release the "refined draft" of the gene regulatory network by solving the proposed mathematical program. To solve large instances of the mathematical program we run the ACO metaheuristic described in §3.4.5.

In the following sections, we first present a computational analysis on randomly generated instances to verify the viability of the proposed algorithm to solve instances of the GRNP problem; then we show the results obtained by applying the proposed methodology to a real instance based on *Saccharomyces cerevisiae* genome data.

3.4.7. Randomly Generated Instances

The computational analysis on random instances was carried out on different instances each of them having 100 nodes. The instances were generated by varying the average *degree* of nodes, from 10 to 12 with a step size of one half. The instances were ordered in 5 groups with increasing graph density. Each group consisted of 5 instances with approximately the same density but with a different seed of the random number generator. The minimum number of neutral nodes was set to 80 in all cases. The groups of instances are reported in Table 3.2 and they are available at *http://homepage.univie.ac.at/martin.romauch/GRNP/*.

The instances we generated are already difficult to solve optimally for a commercial solver

3.4. Inference of Gene Regulatory Networks

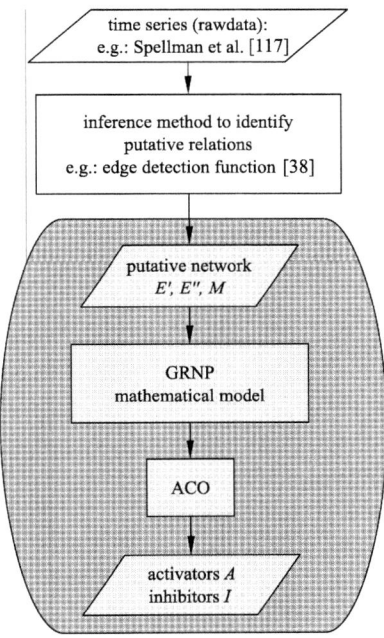

Figure 3.16.: Methodology

Table 3.2.: Five groups of instances with increasing graph densities

$degree$	$seed1$	$seed2$	$seed3$	$seed4$	seed5
10	100-20-1.dat	100-20-2.dat	100-20-3.dat	100-20-4.dat	100-20-5.dat
10.5	100-21-1.dat	100-21-2.dat	100-21-3.dat	100-21-4.dat	100-21-5.dat
11	100-22-1.dat	100-22-2.dat	100-22-3.dat	100-22-4.dat	100-22-5.dat
11.5	100-23-1.dat	100-23-2.dat	100-23-3.dat	100-23-4.dat	100-23-5.dat
12	100-24-1.dat	100-24-2.dat	100-24-3.dat	100-24-4.dat	100-24-5.dat

3. The Double Set Cover Problem

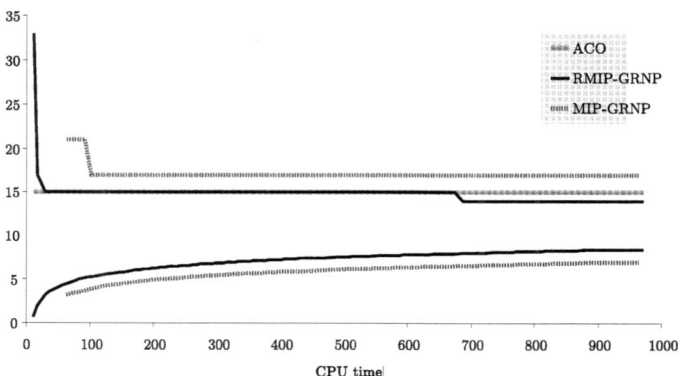

Figure 3.17.: CPLEX solution process for both [*GRNP-MIP*] and [*RMIP-GRNP*], on instance 100-21-4.dat

like CPLEX 10.0. In our computational tests, a time limit of 5000 seconds was imposed for CPLEX. CPLEX was able to compute the optimal solution within the time limit in only two cases out of 25. While solving an instance, CPLEX also returned a lower bound on the optimal solution. The quality of lower bounds was poor and the gap between the lower bound and the solution remained quite large, even after 1 day of computation on 1GHZ machine.

Figure 3.17 displays the trajectories of the solution value and of the lower bound during the execution of branch-and-bound algorithm with grey dashed lines. On the same diagram, the trajectory of the ACO solution is reported with a grey solid line, while black solid lines represent the trajectories of the solution value and of the lower bound during the execution of the branch-and-bound algorithm applied to [*RMIP-GRNP*]. [*RMIP-GRNP*] is a relaxed formulation of [*GRNP-MIP*] obtained by relaxing constraints (3.40) and (3.41). It allows us to compute better lower bounds of the solutions. In fact, the lower bound of [*RMIP-GRNP*] is itself a lower bound for [*GRNP-MIP*].

[*RMIP-GRNP*] is easier to solve for CPLEX. Within a time limit of 1000 seconds, CPLEX optimally solved 6 instances of [*RMIP*] while it did not solve any instance of [*GRNP-MIP*] in the same amount of time. In only two cases, the lower bound obtained from [*RMIP-GRNP*] was worse than the one returned by CPLEX when solving [*GRNP-MIP*], whereas it was significantly better in more than half of the instances as reported in the last two columns of Table 3.3. From Figure 3.17 it is evident that [*RMIP-GRNP*] converges faster and gives better lower bounds especially in early stage. Note that, the trajectories depicted in Figure 3.17 do not refer to a peculiar case (instance *100-21-4.dat*). Indeed, all the instances exhibited similar trajectories.

3.4. Inference of Gene Regulatory Networks

In the last four columns of Table 3.3, the results on the best solution computed by CPLEX within the time limit and the corresponding lower bounds are reported.

To compute good quality solutions of large instances, we implemented the ACO procedure described in §3.4.5. We also compared ACO with GRASP and GREEDY. GRASP, Greedy Randomized Adaptive Search Procedures, is another metaheuristic methodology to solve combinatorial optimization problems, see [36, 37, 99]. Its basic idea is to disturb a greedy construction heuristic and it can also be considered as a simplification of ACO, that is, an ACO without common memory (pheromone). Hence, by comparing ACO and GRASP we can verify the effectiveness of pheromone. GREEDY is a greedy heuristic improved by a local search. The local search procedure implements a 2-exchange improvement between activator, inhibitor and neutral nodes.

We imposed a time limit of 180 seconds for both ACO and GRASP and we started the two algorithms with 5 different seeds since they are random search procedures. Table 3.3 summarizes the computational results. For each instance, the value of the best solution computed within the time limit by each procedure is listed. For ACO and GRASP, the average value computed on all the solutions constructed during the execution of the algorithm is also reported within brackets.

In all the instances, the ACO metaheuristic outperforms significantly GREEDY and GRASP. Indeed, the performances of the greedy heuristic are poor. The algorithm often gets trapped in infeasible solutions and the local search procedure is mostly devoted to restore the feasibility of solutions. The capability of GREEDY to compute good quality solutions is thus compromised. By comparing ACO and GRASP, we deduce that the effect of pheromone is considerable. In 24 instances out of 25, ACO computed solutions which are not worse than those computed by CPLEX in 5000 sec., and in 19 cases ACO outperforms CPLEX in terms of solution quality. Only for instance 100-22-5.dat, CPLEX provided a solution which was better than the ACO solution.

In summary, ACO gave high quality solutions for the GRNP problem in a reasonable amount of time, and we may conclude that it is capable of success.

3.4.8. A Real World Problem

To show the viability of the proposed approach we considered a real test based on microarray experiments on Saccharomyces cerevisiae. More specifically, we considered a 17-point time series data set measuring the expression level of each of 6601 different genes from the S. cerevisiae genome, available at *http://genome-www.stanford.edu/cellcycle/ /data/rawdata/*.

To create the putative network we implemented the improved edge detection function pro-

3. The Double Set Cover Problem

Table 3.3.: Computational results for the GRNP instances

	GREEDY	GRASP	ACO	CPLEX	CPLEX	LB (GRNP-MIP)	LB (RMIP-GRNP)
time limit (sec)		180	180	5000	1000	1000	1000
100-20-1.dat	32	22(23.6)	21(21)	21	21	15	17
100-20-2.dat	35	18(19)	14(14.2)	15	-	13	10
100-20-3.dat	26	16(18.2)	13(13.8)	13*	16	10	10
100-20-4.dat	28	20(21.2)	17(17)	18	18	12	16 [†]
100-20-5.dat	47	19(20)	17(17.2)	17*	18	15	15 [†]
100-21-1.dat	40	19(21.2)	16(16.6)	17	17	11	13 [†]
100-21-2.dat	24	14(15.2)	10(10.2)	10	10	3	5
100-21-3.dat	20	15(16.4)	11(11.2)	13	13	3	5
100-21-4.dat	30	16(18.2)	15(15.2)	16	17	7	9
100-21-5.dat	29	15(18.6)	15(15)	15	15	14	13 [†]
100-22-1.dat	26	15(16.8)	13(13.2)	14	14	6	11 [†]
100-22-2.dat	26	12(14)	9(9.2)	11	11	2	3
100-22-3.dat	22	12(14)	8(8.4)	13	13	3	3
100-22-4.dat	20	14(16)	11(11.8)	13	18	3	4
100-22-5.dat	21	14(16)	13(13)	12	12	10	11 [†]
100-23-1.dat	22	13(13.6)	10(10)	11	11	3	5
100-23-2.dat	15	10(11.4)	6(6)	9	9	1	1
100-23-3.dat	21	10(11.2)	6(6.6)	8	10	2	3
100-23-4.dat	18	10(12.8)	9(9)	10	12	2	3
100-23-5.dat	18	13(15)	10(10.4)	12	12	3	4
100-24-1.dat	19	12(12.8)	9(9)	10	10	2	2
100-24-2.dat	27	7(8.4)	4(4.2)	8	8	0	0
100-24-3.dat	10	6(7.6)	4(4)	4	5	1	1
100-24-4.dat	16	10(11.4)	5(5.2)	8	9	1	1
100-24-5.dat	15	11(13.2)	7(8.2)	9	12	3	3

* marks CPLEX optimal solutions.
[†] marks optimal solutions of *RMIP-GRNP*.

3.4. Inference of Gene Regulatory Networks

posed in [38]. This methodology suggests a relation between two genes as a result of a local analysis on the expression time series of the two genes. More in detail, it investigates local changes in the expression data and their qualitative behavior, e.g. increasing or decreasing, in order to discover either resemblances or differences. The edge detection function scores any pair of genes $(g_a, g_b \in N)$ with a value $S(g_a, g_b) \in [-1, 1]$. Once all the possible pairs of genes have been scored, the corresponding influences are considered in the putative regulatory network if the absolute value of their score is greater than a given threshold value (μ), that is, $|S(g_a, g_b)| \geq \mu$.

In our computational analysis we considered several values of μ. The higher was the threshold value μ, the smaller was the number of influences detected by the edge detection function method. For values of $\mu \geq 0.5$, the putative network was rather small with many false negative influences. In these cases, it was not convenient to apply our procedure. With a threshold value equal to 0.3 ($\mu = 0.3$), the nodes of the putative network had an average degree of 14. Since there were nodes without putative inhibition/activation we added the auxiliary nodes AUX_1, AUX_2, and the corresponding putative regulations. Node AUX_1 activated all nodes with no putative activations and node AUX_2 inhibited all the nodes with no putative inhibitions.

Given the putative gene regulatory network, we generated several instances of the GRNP problem by varying the lower bound on the number of neutral genes, i.e., the parameter M. We considered values of M greater than 90% of the total number of genes. Such values of M are coherent with the number of known regulators, which is approximately 8% of the total number of genes (http://www.yeastgenome.org). To compute the gene regulatory network we run the Ant Colony Optimization metaheuristic. The objective value of the final solution, which corresponds to the average number of irregular influences in the refined gene regulatory network, was on average 523. The average value was computed over all the instances generated with different combinations of M and μ. Moreover, the number of genes labeled in the solution was in general not constrained by $|N| - M$. Indeed, the constraint on the maximum number of regulators, both activators and inhibitors, was active only for values of $M \geq 98\%$.

These results, i.e., the relatively small number of regulators and the high number of irregular influences, implicitly corroborate the conclusion that the putative network has a scale-free structure and is preserved by the method we proposed.

Figure 3.18 shows a small portion of the refined network computed with the following settings: $M = 95\%$ and $\mu = 0.3$. This portion is extracted by choosing one inhibitor (node 394, YBR293w) and its regulating and regulated neighbors. Activating and inhibiting influences are represented by arrows and T-ending arcs respectively. The nodes are labeled with numbers and the key to the corresponding genes is given in Table 3.4.

As displayed in Figure 3.18, the influences proposed by our approach have been scored

3. The Double Set Cover Problem

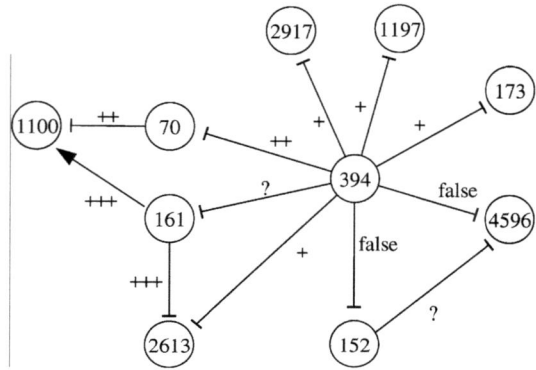

Figure 3.18.: A small cut-out of the resulting network

Table 3.4.: Node label and the corresponding gene product

label	gene	label	gene
70	YAL005c/SSA1	2917	YLR293c/GSP1
152	YBL025w/RRN10	1100	YDR524c
161	YBL016w/FUS3	1197	THR3
173	YBR001c/NTH2	2613	YKR065c
394	YBR293w	4596	YNL162w/RPL41A_ex1_f

3.4. Inference of Gene Regulatory Networks

according to the biological knowledge up to date [21]. With "+" we tag all the relationships between genes which are consistent with the current knowledge about the S. cervisae genome. The number of "+" denotes the probability of the existence of a biological and functional connection. In particular, the arcs tagged with "+++" are deeply documented by scientific biological papers, such as the influence of 161 (YBL016w/FUS3) upon 1100 (YDR524c) and 2613 (YKR065c), [97]. Fus3 is a kinase involved in cell proliferation processes and it also acts in a pathway upstream of Arf (aka Age1 number 1100, YDR524c) which is strongly implied in cell proliferation processes as well, [126].

With only one "+" we tagged influences which are consistent with a gene ontology analysis, i.e., they may be inferred from other genes which belong to the same cluster and exert similar biological functions. For instance, referring to the relationship between 394 and 1197, a defect in the uptake of histidine, lysine, or arginine is also observed in the vacuolar membrane vesicles of mutants YBR293w (VBA2). More specifically, VBA2 (a vesicle amino acid transporter) is involved in amino acid uptake from the external environment (as shown by mutants [114]). It is known that histidine and arginine biosynthesis genes, His1 and Arg6, are inhibited by a feedback regulatory loop when amino acids are available in the cell [76]. Therefore, perturbing the activity of VBA2, it is possible to modify intracellular amino acid levels, and thus His1 and Arg6 (through the regulatory loop). With regard to Thr3, it is in a gene cluster together with His1 and Arg6, therefore it is highly probable that Thr3 is involved in amino acid biosynthesis as well, and is regulated by the same feedback loop. That may provide a rationale for Thr3 ⊢ VBA2 relation found.

Our methodology also revealed influences that at the moment are unknown, tagged with "?". This is due to either one or both of the two following motivations.

1. the relation between the two proteins is unknown.

2. the proteins do not have a known functionality.

Finally, we tagged with "false" those influences which correspond to false positive, since between the two genes there is not any functional and/or physical relation.

The relations may be a good initial point for discovering interesting regulatory coherence and may be the basis for designing experiments.

3.4.9. Conclusion

In this section, we presented a mathematical model which formalizes the problem of identifying a smaller set of interesting candidate regulatory elements given a putative gene regulatory

3. The Double Set Cover Problem

network. The Ant Colony Optimization procedure we proposed to solve real instances of the problem, provided good quality solutions in a reasonable amount of time.

The proposed methodology, like any reverse engineering approach, might not be lacking of errors, since both false positive and false negative influences might be revealed. Indeed, data are often inherently inadequate to identify the gene regulatory network. For instance, simple signal analysis techniques fail to find the vast majority of known regulatory relations on the Cho/Spellman data set as demonstrated in [38]. Better results could be achieved if more recent methodologies were applied in the generation of the putative gene regulatory network. On this subject, we can mentioned among others, methods which either use disparate biological data sources [9], use multiple time series data sets [113] or relieve from biological data errors [64].

The proposed approach is fairly general and makes interesting prediction of the S. cerevisiae gene regulatory network. However further issues should be addressed. For instance, gene products are also activated by other factors such as phosphorylation, an option which is not considered in our model.

We believe that the proposed method should be used in combination with other methodologies and it could serve as a basis to design experiments with the aim of discovering unknown influences.

3.5. Art Gallery Problems

In this section we give complexity results for the *Vertex Guard Double Cover* problem and give some notes on how to get solutions and also give bounds for this problem.

Definition *Vertex Guard Double Cover Problem*
The layout of art gallery is given by a polygon $\mathcal{P}$ described by the graph $G(V, E)$ where V is also called the set of corner points. A *Vertex Guard Double Cover* is defined by:

- a pair of sets (Λ_1, Λ_2) and $\Lambda_1, \Lambda_2 \subset V$.

- $\Lambda_1 \cap \Lambda_2 = \phi$.

- Λ_1 is a Vertex Guard Cover of $\mathcal{P}$.

- Λ_2 is a Vertex Guard Cover of $\mathcal{P}$.

To interpret this problem we can imagine to position cameras on finite number of possible points (not necessarily corners) of a polygonal region. The aim is to monitoring the border of the polygon using two independent systems. For instance we can imagine that the camera

3.5. Art Gallery Problems

systems are sensible to two different kinds of radiation. Furthermore we don't allow to put more than one camera into one and the same place.

We can refine the problem and use weights instead of solely counting the number of cameras:

Definition *Minimum Cost Vertex Guard Double Cover Problem*
Again the layout of art gallery is given by a polygon $\mathcal{P}$ with the corresponding graph $G(V, E)$. The pair (Λ_1, Λ_2) is a solution to the *Minimum Cost Vertex Guard Double Cover Problem* if:

- $\Lambda_1, \Lambda_2 \subset V$.
- $\Lambda_1 \cap \Lambda_2 = \phi$.
- Λ_1 is a Vertex Guard Cover of $\mathcal{P}$.
- Λ_2 is a Vertex Guard Cover of $\mathcal{P}$.
- $c^1 : V \rightarrow \mathbb{R}^+$
- $c^2 : V \rightarrow \mathbb{R}^+$
- and if $\sum_{i \in \Lambda_1} c^1(i) + \sum_{i \in \Lambda_2} c^2(i)$ is minimal.

3.5.1. Complexity Issues

The question about complexity is not covered by the results in the previous section. Although the problem is a version of the DSCP we find a very special structure. I.e.: $A_1 = A_2$ and the sets S_i describe intersection of the visibility polygon of vertex i with the border of the polygon.

Theorem 3.5.1. *The VGDCP (with holes) is NP-hard.*

Proof. A sketch of the proof.
Assumption: $\mathcal{P}$ may contain holes. We give a proof where we suppose that the graph is cubic. A sketch: we transform the vertex cover on cubic graphs $G(V, E)$. Given an instance we transform the network to corridors $\mathcal{P}(V, E)$, where at the position of the vertices we get a junction of corridors where at lest 2 corner points cover the 3 corridors.
For each junction of corridors we have to identify at least 2 corner points that carry the property of the former vertex. Namely, if a vertex v_i was adjacent to the edge e_j then the corridor c_j plus junction is visible to at least two corner points in the junction J_i. We have to consider three different junction types:

1. no acute angle

3. The Double Set Cover Problem

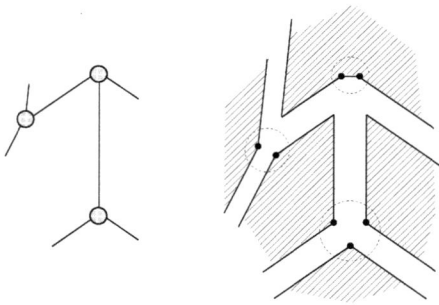

Figure 3.19.: Junction types

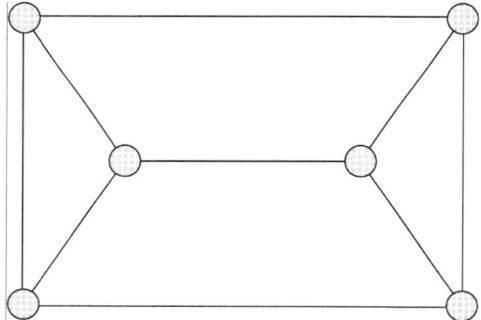

Figure 3.20.: Cubic graph

2. one acute angle

3. two acute angles

In Figure 3.19 we show how to do the transformation. Since the explained method only works for planar graphs we need to deal with edge crossings. Therefore we add special junctions like depicted in Figure 3.22. The idea is that we add alcoves that need to be illuminated by at least one interior light, therefore the outer nodes need not be used in the optimal solution, since the corresponding visibility polygons only cover one of the corridors. E.g. we can see the transformation of a cubic graph (Figure 3.20) to an art gallery instance, depicted in Figure 3.21.

To complete the proof we summarize that if we are given a solution of a vertex cover problem

3.5. Art Gallery Problems

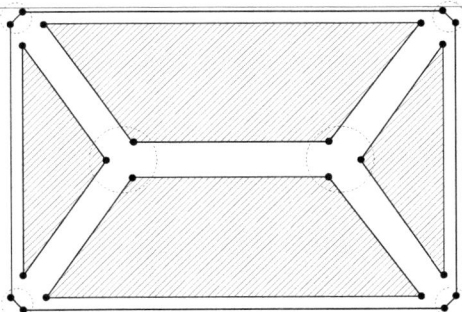

Figure 3.21.: Polygon

with k nodes, then we may position 2 different kinds of cameras in the junctions and therefore we get a solution with $k' = 2k$. On the other hand, if we end up in a solution of the VGDCP that uses $2k$ cameras, then we can argue that we can transform the solution in such a way that the cameras used, are always located in the junctions. If this is true then we can also argue that we will have at most one camera of each type in the same junction and that these cameras cover all adjacent corridors. Otherwise we can eliminate one of them or reposition it. After this step we still don't use more than $2k$ vertex guards and therefore at least on of the vertex guard types doesn't use more than k guards. The associated corridors build the solution for the vertex cover problem. □

Remark 3.5.2. Since Garey and Johnson show that vertex cover problem in planar graphs with maximum degree 3 is still NP-complete in [45], we can get rid of the junctions and we can simplify the proof. For the sake of completeness we show a method how we can transform non-planar graphs by adding special corridor crossings. In 3.22 we can find a non planar graph and we will show how to transform it to an art gallery instance (Figure 3.23) with corridors and alcoves. Figure 3.24 shows the alcoves in detail. We can see that we have to locate 2 cameras inside the alcove and therefore we can argue that it is always possible to find optimal solutions where we don't need to select points from junctions of corridors to cover the corridor.

Theorem 3.5.3. *The VGDCP without holes is NP-hard.*

Proof. We will give a sketch of the proof and concentrate on how to adapt the proof in [81], where 3-SAT is transformed to the vertex guard problem (see Figure 3.25). We recall that fixing a vertex guard means choosing a visibility polygon and how the original proof makes use of alcoves that are only visible to some vertices. In the instance we find narrow corridors

3. The Double Set Cover Problem

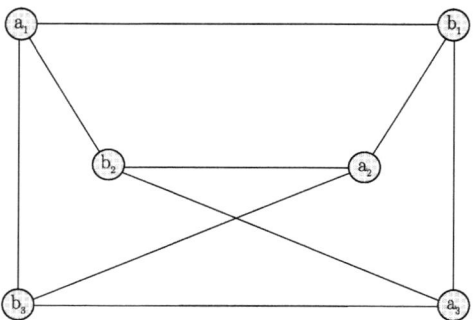

Figure 3.22.: $C_{3,3}$ a non-planar graph

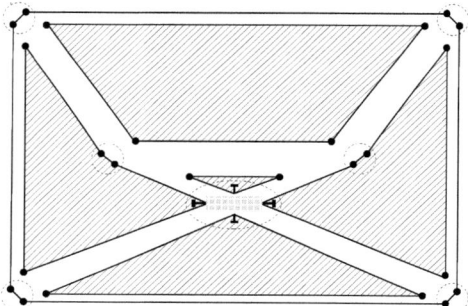

Figure 3.23.: Junction of corridors

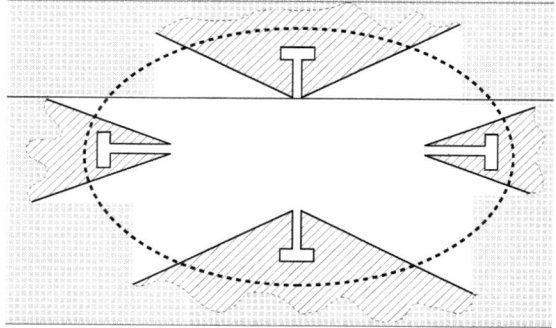

Figure 3.24.: Junction in detail

3.5. Art Gallery Problems

that can be covered by only few vertices. We show how the clauses are transformed for [81] and we show how to transfer this idea. In Figure 3.26 we can see that the black node is a dominating node, it covers all the pikes and the whole surrounding. The white nodes represent the second type of guards. We can see that it is necessary to select at least three white nodes for the second vertex guard type. Furthermore it is not reasonable to put all the nodes on inside positions of the pikes. I.e.: at least one of the white nodes needs to take a bordering position to the exterior. The whole configuration represents a clause and each pike represents a literal. If we mark the outer vertex, then this represents a "true", and the inside choice represents a "no". The connection of the literal variables is realized by narrow corridors that are exclusively visible from the corresponding literal.

The key idea is to redesign the transformation of [81] in a way that one of the vertex guard types uses obvious locations (w.l.o.g the black one), while the other one solves the 3-SAT instance.

To do that we add alcoves to the narrow corridors, i.e.: we add an alcove with a pike, see Figure 3.27. Here the idea is that some parts of the alcove(i.e: the pike) are visible from some exterior points, e.g.: the corresponding node that realizes the literal. We know that the alcove forces us to put at least 2 guards of different types into it. Or to be more precise: if there is no help coming from the exterior we need at least 3 guards. Since the alcove is asymmetric it is clear that if we get assistance from outside then this has a direct influence of the inside alcove configuration. I.e. if the assistance comes from a white node then the black node is the unique position that covers the whole alcove. Obviously the vis-à-vis node gets colored white. In the Figure 3.28 we can see the transformed instance. Investigating the picture we can see that the black nodes give the smallest set of vertex guards.

We start with the upper left corner and it is a good choice to use them for both vertex types (they correspond to the node called W in Figure 3.25 of the proof in [81]). W.l.o.g. we choose them like in the Figure 3.28. Now we continue and have look into the alcoves that represent the clauses. It is obvious to choose the black node that covers the whole alcove and that therefore the white ones have to solve the 3-SAT. The large alcoves in the low left corner represent the values of the variables and there are two alcoves for each variable. One represents "true" and the other one "false". All occurrences are aggregated in the narrow corridors and we can choose which one gets assisted by nodes of literals and which one gets assisted by the nodes for the variables. The details can be found in the proof of [81]. □

3. The Double Set Cover Problem

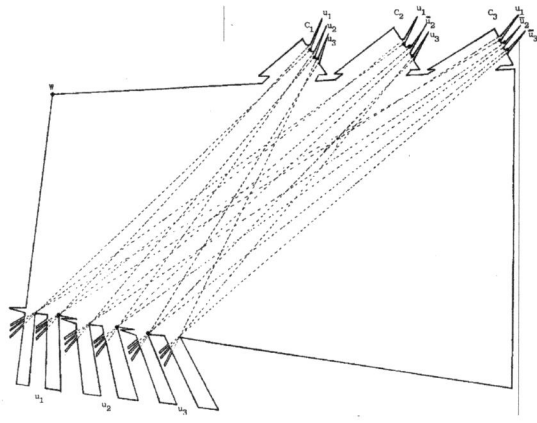

Figure 3.25.: Transformation of 3-SAT to vertex guard [81]

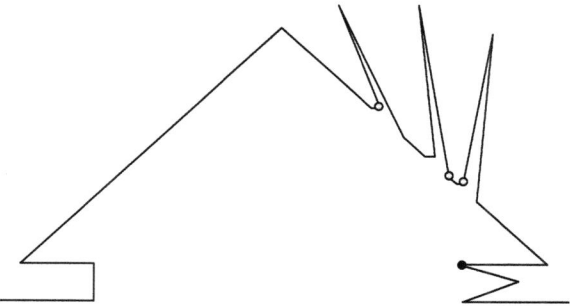

Figure 3.26.: Representation of a clause: the white nodes represent the values of the literals

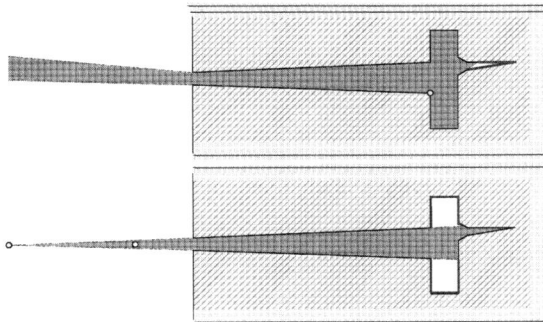

Figure 3.27.: Narrow corridor with an alcove and spike

3.5.2. Bounds

Theorem 3.5.4. *If we we restrict on polygons without holes then the minimum number of vertices needed to solve the* Vertex Guard Double Cover Problem *is bounded by* $\lfloor \frac{2n}{3} \rfloor$.

Proof. The arguments of the proof for Chvátal's classic art Gallery Problem by Fisk [40] are used and an example shows that the bound is sharp. First a triangulation of the polygon is generated and then we assign colors to the vertices in a way that each edge connects different colors and it is shown that only 3 colors are needed (tree structure). Therefore putting together all vertices of one color we get a vertex guard cover for the polygon. To get a solution for a double covering it is sufficient to take 2 colors. Since the number of vertices of least one of the colors needs to be larger or equal $\lceil \frac{n}{3} \rceil$. We can see that the minimum number of two colors is bounded by $\lfloor \frac{2n}{3} \rfloor$. An example where the minimum equals $\lfloor \frac{2n}{3} \rfloor$ is given in the example depicted in Figure 3.29. □

Definition *Neighboring Holes*

Two holes of a polygon $\mathcal{P}$ are neighboring holes if they have an edge that is visible to each other. If $\mathcal{P}$ has h holes that are connected trough a tree of neighboring holes we say that $\mathcal{P}$ has h neighboring holes. We emphasize that we alsosuppose that the corresonding tree is planar and the corresponding pairs of visible edges are disjoint.

Theorem 3.5.5. *If the polygon $\mathcal{P}$ has h neighboring holes then the minimum number of vertices needed to solve* Art Gallery Double Covering *problem is bounded by* $\lfloor \frac{n}{2} \rfloor + \lfloor \frac{n}{3} \rfloor + 1$.

Proof. Sketch of the proof by induction in h: Suppose that $\lfloor \frac{n}{2} \rfloor + \lfloor \frac{n}{3} \rfloor + 4h$ guards are sufficient to cover a polygon with h neighboring holes. Obviously, for $h = 0$ the statement is true. Now

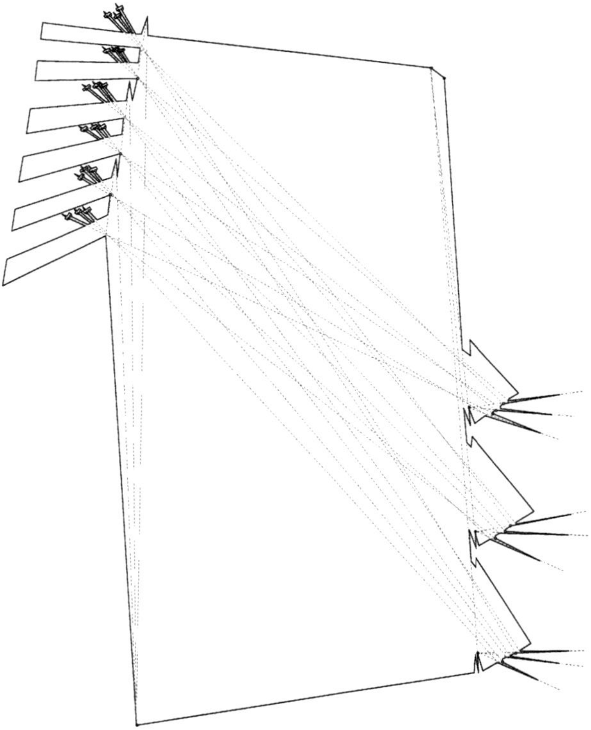

3. The Double Set Cover Problem

Figure 3.28.: Transformation of 3-SAT to the Vertex Guard Double Cover Problem

90

3.5. Art Gallery Problems

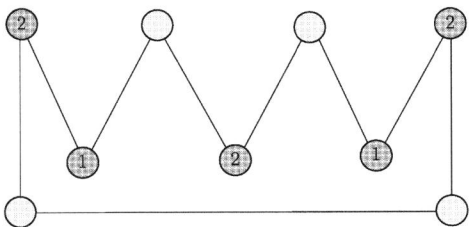

Figure 3.29.: (Chvátal's Comb): an example where the bound is sharp

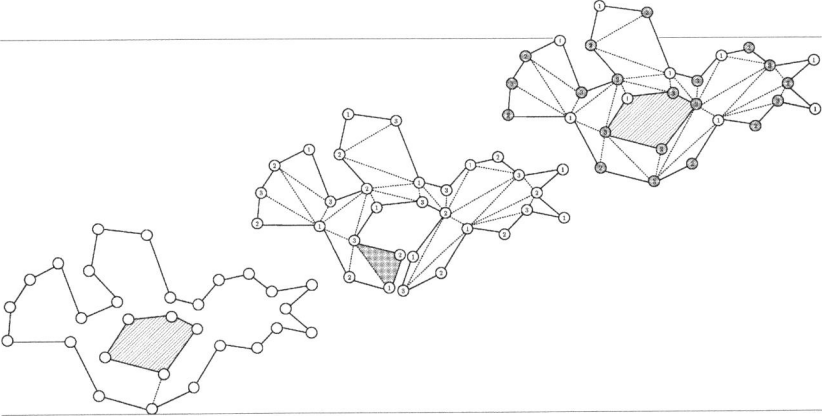

Figure 3.30.: A polygon with one hole gets transformed into a polygon without holes to select the guard sets

we show that the statement is true for $h = 1$: I.e.: we will show that $\lfloor \frac{n}{2} \rfloor + \lfloor \frac{n}{3} \rfloor$ are sufficient to cover a polygon with 1 hole.

We use the property that for each polygon $\mathcal{P}$ with one hole we can find an edge $[u, v] \in E$ of the border of the hole that is fully visible by a vertex $w \in V$ of the border of the polygon or we can find an edge $[u, v] \in E$ of the border of the polygon that is visible to a corner point $w \in V$ of the border of the hole. We cut the polygon $\mathcal{P}$ along the edge $[v, w]$. I.e.: we generate a new polygon $\tilde{\mathcal{P}}$ without holes and $n + 2$ vertices $\tilde{V} = V \cup \tilde{v}, \tilde{w}$ and the edges $\tilde{E}$. We start a triangulation by cutting the ear (w.l.og.: $\{u, v, w\}$). The remaining polygon is arbitrarily decomposed into triangles of diagonals. Figure 3.30 gives an example how the procedure works. After that we assign colors to the nodes and we may encounter the following situations:

3. The Double Set Cover Problem

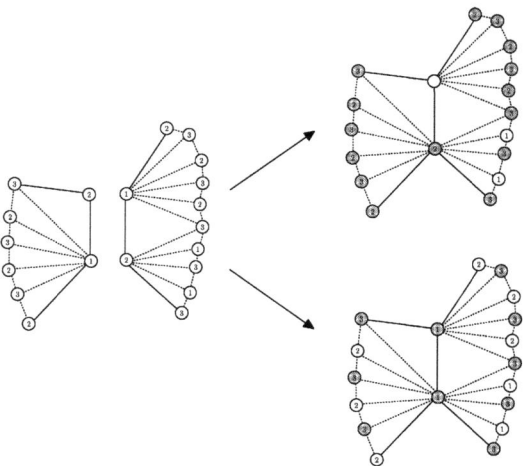

Figure 3.31.: Gluing a cut: we have 2 possibilities to choose the second color

- If $v, w, \tilde{v}, \tilde{w}$ have three different colors then it is possible that v $\tilde{v}$ or w and $\tilde{w}$ have the same color.

- $v, w, \tilde{v}, \tilde{w}$ have two different colors. If the points $\tilde{v}$ and v have the same color then $\tilde{w}$ and w also have the same color. Else $\tilde{v}$ and w have the same color.

We can show that in any case we can use this solution to assign colors to the original polygon $\mathbb{P}$ in a way that 2 colors cover the polygon. Additionally we use the fact that the triangulation of $\widetilde{\mathcal{P}}$ may only force us to keep one of the colors and allows us to choose the second one. For one of the cases we show how this is done. Figure 3.31 gives a schematic procedure. We emphasize that there are two possibilities to choose the second color. We summarize that we might have to choose the color with the maximum number of members, which is not larger than $\lfloor \frac{n}{2} \rfloor + 1$. Regarding the second color, we are free to select the color with the minimum number of members. Therefore we have $\lfloor \frac{n}{2} \rfloor + 1 + \lfloor \frac{n+2}{3} \rfloor$ guards in $\widetilde{\mathcal{P}}$. After the gluing step we may only loose members, therefore the minimum number of guards needed to cover a polygon with one hole is not larger than:
$$\frac{n}{2} + 1 + \lfloor \frac{n+2}{3} \rfloor$$

Now we argue that this is also true for polygons with h neighboring holes. Suppose that $\mathcal{P}$ has $h+1$ holes then we have at least one leaf. We can connect the leaf-hole with the parent-hole by connecting the pair or visible edges. Since this doesn't increase the number of edges we can

3.5. Art Gallery Problems

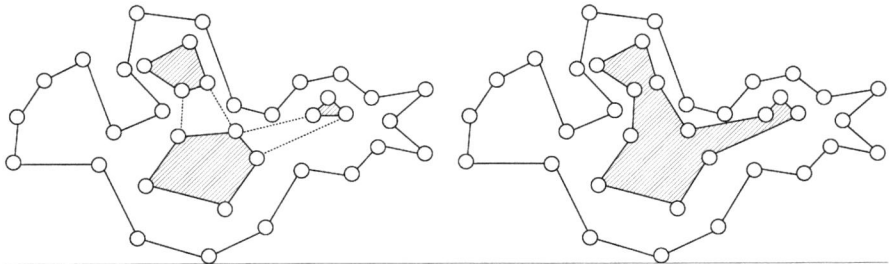

Figure 3.32.: Polygon with three neighboring holes can be transformed to a polygons with one hole

use the assumption for h holes which leads us to the result. In Figure 3.32 we can see an example. □

3.5.3. Solution Techniques

Guard placements algorithms often make use of triangulations. Following Fisk's proof [40] of Chvátal's theorem for art gallery problems, we can see the important role of dividing a polygon into triangles by diagonals. Finding the number of such divisions for convex polygons it is called *Euler's Polygon Division Problem*. And the answer is given by the Catalan's numbers C_n, i.e.: C_n is the number of triangulations of a convex polygon with $n+2$ vertices.

$$C_n = \frac{1}{n+1}\binom{2n}{n} = \frac{(2n)!}{(n+1)!\,n!} \qquad \forall n \geq 0.$$

Stanley [118] gives various other interpretations of C_n. E.g: the number of monotonic paths in a (n,n)-grid that do not cross the diagonal (equivalent to *Dyck Words*). An illustration is given in Figure 3.33: we depicted $C_3 = 5$ triangulations of the pentagon and how they can be interpreted as binary trees with 4 leaves, represented by putting 3 pairs of parentheses for 4 objects.

It might be intuitive, that for solving the minimum vertex guard problem it is sufficient to investigate all possible triangulations and picking the "color" with the smallest number of members. C_{n-2} gives an idea of the size of the corresponding solution space. The following example (Figure 3.34) shows that the solutions space is not sufficient to reach all possible solutions. The key problem of the method is that only using diagonals permits the guards to cross their views. In fact, it is possible to construct examples that keep the triangulation solutions arbitrarily far away from optimum (see Figure 3.35).

3. The Double Set Cover Problem

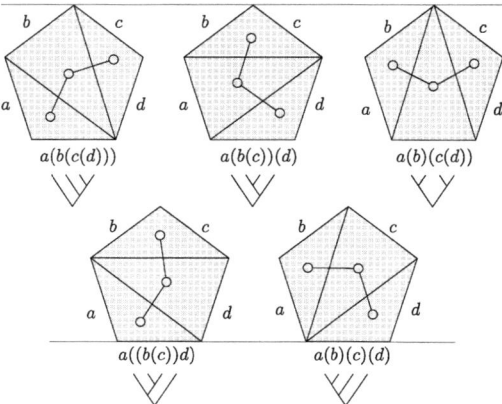

Figure 3.33.: Catalan numbers

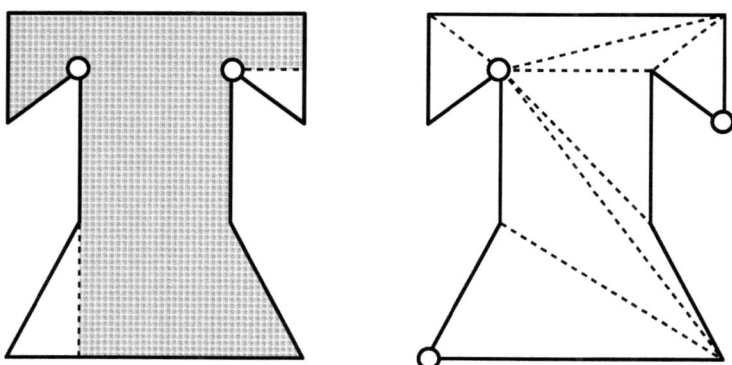

Figure 3.34.: Comparing the optimal solution with the best solution reachable by the triangulation technique

3.5. Art Gallery Problems

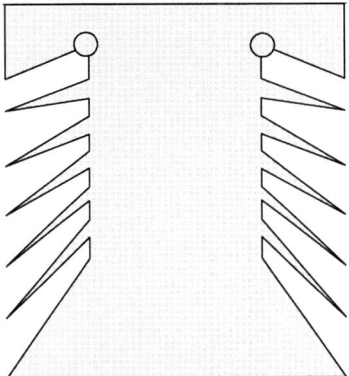

Figure 3.35.: The optimal solution uses two vertices, while the number of vertices needed by solutions that correspond to triangulations is dependent on the number of spikes

A. Notions from Graph Theory and Computational Geometry

A.1. Graph Theory

The definitions that follow can be found in introductory books in graph theory. E.g.:[25]. Most of definitions only refer to directed graphs (digraphs).

Definition *Degree*
the degree of a vertex in an undirected graph $G(V, E)$ is the size of its neighborhood, i.e.: $d(i) = |\{j : \{i,j\} \in E\}|$. The in-degree and the out-degree of a vertex in a digraph is number of predecessors and successors, and the degree of a graph is the maximum degree of all vertices. I.e:

$$d^+(i) = |\{j : [i,j] \in E\}| \qquad d^+(G) = \max_{i \in V}\{d^+(i)\}$$
$$d^-(i) = |\{j : [j,i] \in E\}| \qquad d^-(G) = \max_{i \in V}\{d^+(i)\}$$
$$d(i) = d^+(i) + d^-(i) \qquad d(G) = \max_{i \in V}\{d(i)\}$$

Definition *Induced Subgraph*
$G'[V'] = G'(E', V') \subset G(V, E)$ is an induced subgraph of G if $i, j \in V'$ implies that $\{i,j\} \in E'$.

Definition *Spanning*
An induced Graph is *spanning* if $V' = V$.

Definition *Complete Graphs*

- The complete graph $K_n = G(V, E)$ has n vertices and for all $v, w \in V$ it follows $v \neq w \Rightarrow [v, w] \in E$. If G is a undirected graph then $v \neq w \Rightarrow \{v, w\} \in E$
- A complete bipartite graph $K_{n,m} = G(V = V_1 \cup V_2, E)$ satisfies $|V_1| = n$ and $|V_2| = m$. If and only if $v \in V_1$ and $w \in V_2$ E contains $\{v, w\}$, respectively $[v, w]$.

A. Notions from Graph Theory and Computational Geometry

Definition *Path*
A directed path P is a sequence of adjacent edges in a graph $G(V,E)$. Suppose that number of edges is L (length of the path) then we can write as $P = \{[v_i, v_{i+1}] \in E : i = 1\ldots L\}$. The path is simple if all v_i are different. For the undirected case is analogously defined.

Definition *Circuit*
A path C is a circuit or cycle if $|\{v_i : i = 1\ldots n\}| = n$.

Definition *Regular*
a undirected graph $G = (V,E)$ is called *k-regular* if every node v in V has degree k in G, i.e. v is incident with precisely k edges.

Definition *Edge Cover*
In undirected graphs, an edge cover is a set of edges $S \subset E$, such that every node in $i \in V$ is incident with at least one edge in S. I.e: $\forall i \in V \exists e \in S : i \in e$.

Definition *Matching*
A matching is a set of edges $M \subset E$, such that different edges $e_1, e_2 \in M$ are node disjoint. If a matching is also spanning then it is called a *perfect matching*.

Definition *Covering and Packing*
Given a class of graphs $\mathcal{H}$, then there the covering problem is to find a minimum number of subgraphs in $G(V,E)$ that are isomorphic to graphs in $\mathcal{H}$, such that the union of the subgraph covers all nodes. The packing problem is to find a maximum number of disjoint subgraphs isomorphic to elements in $\mathcal{H}$.

Definition *Tree*
A graph $G(V,E)$ is a spanning tree if spanning and doesn't contain cycles. As a consequence E has $|V| - 1$ edges.

Definition *k-Factor*
k-factor A *k-factor* is a *k*-regular spanning subgraph. E.g.: 1-factors are also known as *perfect matchings* (i.e. matchings which cover all nodes), and a 2-factor is a spanning node-disjoint circuit.

Definition *factorization*
A *factorization* is a partition of a graph into connected spanning subgraphs. A k-factorization is a factorization into k-factors. E.g: a 1-factorization is partition into matchings.

A.2. Computational Geometry

Definition *Polygon*
A simple polygon $\mathcal{P}$ without holes in the plane can be described as a undirected graph $G(V.E)$. Where V are the corner points of the polygon and E is a cycle that describes the border of the polygon. Every vertex $i \in V$ corresponds to a point $P_i \in \mathbb{R} \times \mathbb{R}$ in the plane. For intersections of two segments we assume that $[P_i, P_{i+1}[\cap]P_j, P_{j+1}[= \phi$. This is sufficient to assume that $\mathcal{P}$ has an interior. Polygons with holes can be defined recursively adding simple polygons without holes into the interior of a "master polygon". Then the interior of the polygonal holes is added to the complement of $\mathcal{P}$. For Art Gallery Problem we us the term *visibility polygon*:

Definition *Visibility Polygon*
Visibility Polygons $\mathcal{V}$ describe the visible region of a point v in the interior or on the boundary of a polygon $\mathcal{P}$. Suppose that a point $v \in \mathcal{P} \subset \mathbb{R}^2$ is given, then the visibility polygon $\mathcal{V}(v, \mathcal{P})$ collects all segments in $\mathcal{P}$ that start in v and continue until they reach a point where we first intersect with the complement $\mathcal{P}^c$.

$$\mathcal{V}(v, \mathcal{P}) = \bigcup_{a \in \mathbb{R}^2 \wedge [v, v+ta] \cap \mathcal{P}^c = \phi} [v, v+ta]$$

Definition *Diagonals*
The connection of two corner point P_i P_j that are not adjacent is called a *diagonal* if $[P_i, P_j]$ is fully contained in $\mathcal{P}$. If parts of the interior of the connection $]P_i, P_j[$ are contained in the boundary, then the diagonal is a *degenerated diagonal*. A pair of diagonals $\{P_i, P_j\}$ $\{P_{i'}, P_{j'}\}$ are called *intersecting diagonals* if $|]P_i, P_j[\cap]P_{i'}, P_{j'}[| = 1$.

Definition *Triangulation into Diagonals*
A *triangulation into diagonals* of a polygon $\mathcal{P}$ with n corner points is a set of $n - 2$ non-intersecting diagonals. Finally a set of $n - 2$ diagonals is called a *degenerate triangulation into diagonals* if a pair of diagonals exists such that $[P_i, P_j] \subset [P_{i'}, P_{j'}]$.

B. NP-complete and NP-hard

B.1. Complexity Theory

Complexity Theory deals with the tractability and the intractability of problems [46]. The basis is the formalization of computation, the corresponding computability and the computational effort. Problems in this formal setting are formal yes/no questions like: Is n a prime number? An input for this problem is $n = 5$ and the corresponding problem instance is given by: Is 15 a prime number? A solution to this problem is a algorithm that terminates with "yes" or "no" for any problem instance. Such an algorithm can be implemented on a Turing Machine TM. The TM represents a very general and abstract computing device, but still there are problems that cannot be be implemented on a TM (e.g. the halting problem). I.e.: there exist problems that are undecidable on TM. We continue with less ambiguous problems and define the class NP of those problems that are solvable on a nondeterministic TM in polynomial time. A NTM is an accelerated TM, where we can think of TMs that work in parallel. This class contains very hard problem like the *Satisfiability Problem* SAT. SAT was the first problem that was shown to be one of the hardest problems in NP. I.e.: S. Cook proved that if an "efficient" algorithm for the SAT is known, then all problems in NP are efficiently solvable. In this context "efficient" means that the effort of computation is bounded by a polynomial. This result motivated to define a set of the most difficult problems in NP - the class of NP-complete problems. Now we turn to "easy" problems. From macro point of view a problem X is easy to solve if the effort to solve an instance $x \in X$ of size $n(x)$ is bounded by a polynomial, then we also say that the corresponding algorithm is polynomial. It is clear, that algorithms with high degrees don't make the problem really easy, therefore this news has no direct practical impact, but it is a good starting point to improve algorithms. Especially section 3 contains various NP-completeness and NP-hardness proofs. The methodology is simple: we transform a known NP-complete problem X to the investigated problem Y, in such a way that solving the the transformed problem answers the primal one. I.e.: Normally, we give an algorithm $\mathcal{A}$ that transforms any instance $x \in X$ to an instance $\mathcal{A}(x) \in Y$ and we show that answering $\mathcal{A}(x)$ is equivalent to answering x. If this algorithm $\mathcal{A}$ is polynomial then $\mathcal{A}$ is a polynomial reduction of X. Since Optimization problems X: $\min\{f(x) : x \in \Omega\}$ are not stated as decision problems, the have

B. NP-complete and NP-hard

to be adopted. We say that an optimization problems is NP-hard if it is as least as hard as a NP-complete problem. I.e.: If we can proof that the corresponding decision problem X_k: $\{f(x) \leq k : x \in \Omega\} \neq \phi$ is NP-complete, then X is NP-hard.

In the following section we present a selection of NP-complete problems, like the 3-SAT that are often used in NP-completeness and NP-hardness to proofs.

B.2. Selected Problems

Definition Satisfiability(SAT).
Given a set $U = \{x_1, \ldots, x_n\}$ of variables and collection C of clauses over U. I.e.: Let $c \in C$ is a clause consiting of literals $c = (l_1 \vee l_2 \vee \ldots \vee l_m)$, where $l_j \in U$ or $\neg l_j \in U$. The collection of clauses C is satisfiable if there exists a truth assignment $f : U \mapsto \{\text{true}, \text{false}\}$ such that $c =$true for all $c \in C$.

Definition *Three-Satisfiability (3-SAT)*
Let U be of variables, collection of clauses over U such that each clause $c \in C$ has at most 3 literals ($|c| \leq 3$) Is there a truth assignment for U that simultaneously satisfies all clauses?

Definition *Three-Exact Cover Problem (X3C)*
X3C is a variant of SCP: Given a finite set $S = \{1 \ldots 3n\}$ and a collection of subsets of S, namely $\mathcal{F} = \{S_i \subset S : |S_i| = 3 \wedge i = 1 \ldots m \}$. Does $\mathcal{F}$ contain a cover of S of size n? Or equivalently, does a subset $\Lambda \subset \{1 \ldots m\}$ of size n exist such that $S = \bigcup_{i \in \Lambda} S_i$?

Definition *Vertex Cover Problem*
In undirected graphs, a *Vertex Cover* is a set of vertices $W \subset V$, such that every edge $e \in E$ is incident with at least one vertex in W. I.e: $\forall e \in E \exists i \in W : i \in e$. The *Vertex Cover Problem* asks for a *Vertex Cover Problem* of size k.

Theorem B.2.1. *Vertex Cover on cubic graphs is NP-complete.*

Remark B.2.2. this is also true for planar cubic graphs

Definition *Vertex Coloring*
Given a graph $G(V, E)$ then $\varphi : V \to \mathbb{N}$ is a k-coloring of the vertices if:

- $e = \{v_1, v_2\} \in E \Rightarrow \varphi(v_1) \neq \varphi(v_2)$
- $|\varphi(V)| = k$

Definition *Edge-Coloring*
Given a graph $G(V, E)$ then $\varphi : E \to \mathbb{N}$ is a k-coloring of the edges if:

- $\varphi(e_1) = \varphi(e_2) \quad \Rightarrow \quad e_1 \cap e_2 = \phi$
- $|\varphi(E)| = k$

Definition *Chromatic-Index*
The chromatic index $\chi_1(G)$ of a graph $G(V, E)$ is the minimum k for which we can find a k-coloring of the edges.

Remark B.2.3. Considering Chromatic-Index for cubic graphs is also called Tait coloring. I.e.: Are 3 colors sufficient to color tie cubic graph G. We can see that a 1-factorization solves Chromatic-Index.

For cubic 2-connected graphs we know the 3 colors are sufficient:

Theorem B.2.4. *(Tait): A bridgeless cubic planar graph can be face-colored with 4 colors if and only if it can be edge-colored with 3 colors.*

Theorem B.2.5. *Chromatic-Index is NP complete. This is still true for cubic graphs [61]:*

Definition *Maximum Two-Satisfiability (Max-2-SAT)*
Given a set U of variables, and a collection of clauses over U such that each clause $c \in C$ has at most two literals $|c| = 2$. Is there a truth assignment for U that simultaneously satisfies at least K of the clauses?

C. Experiments and Randomly Generated Instances

C.1. Randomly Generated DCP Instances

In the experiments on randomly generated instances for the ODCP1 (3.7) and ODCP2 (3.16) (see section 3) we fixed $m = n$ (network structure) and varied the density. The algorithm construct two independent $0-1$ square matrices for the service A_1 and A_2. The entries are uniformly distributed with the exception of sparse matrices where we avoid empty rows. The procedure works as described in the the pseudo code 4. We note that $\frac{M}{N}$ is the the probability that a one is regularly assigned to an entry. If the whole row is zero then we randomly choose exactly one position with a nonzero entry. We calculate the average number n_0 of situations when we encountered an empty row and we abbreviate $d = \frac{M}{N}$:

$$n_0 = n(1-d)^{n-1}$$

To calculate the expected value of the density d' we have to take n_0 into account:

$$d' = f(d,n) = \frac{n(n-1)d + n_0}{n^2} = d - \frac{d - (1-d)^{n-1}}{n}$$

In Table C.1 we can see the relative error $\frac{d'}{d} = \frac{f(d,n)}{d}$ and in Table C.2 we can see $d = f^{-1}(d', n)$ for some values d' and n. For larger values of n we can investigate an asymptotic behavior:

$$\lim_{n \to \infty} (d - d') = \lim_{n \to \infty} \left(\frac{d - (1-d)^{n-1}}{n} \right) = 0$$

C. Experiments and Randomly Generated Instances

Algorithm 4 Generating random instances

$A_1 \leftarrow 0_{n,n}$;
$A_2 \leftarrow 0_{n,n}$;
for $i \in \{1 \ldots n\}$ **do**
    for $j \in \{1 \ldots, i, i+1, \ldots n\}$ **do**
        pick two random numbers $z_1, z_2 \in \{1, \ldots, N\}$;
        if $z_1 \leq M$ **then**
            $A_1(i,j) \leftarrow 1$;
        end if
        if $z_2 \leq M$ **then**
            $A_1(i,j) \leftarrow 1$;
        end if
    end for
    if $\forall j : A_1(i,j) = 0$ **then**
        pick a random numbers $z \in \{1, \ldots, n-1\}$;
        if $z < i$ **then**
            $A_1(i,z) \leftarrow 1$
        else
            $A_1(i,z+1) \leftarrow 1$
        end if
    end if
    if $\forall j : A_2(i,j) = 0$ **then**
        pick a random numbers $z \in \{1, \ldots, n-1\}$;
        if $z < i$ **then**
            $A_2(i,z) \leftarrow 1$
        else
            $A_2(i,z+1) \leftarrow 1$
        end if
    end if
end for
return v;

C.1. Randomly Generated DCP Instances

Table C.1.: Relative error $\frac{d-d'}{d}$

	10%	20%	30%	40%	50%	60%	70%	80%	90%
10	-28.74%	3.29%	8.65%	9.75%	9.96%	10.00%	10.00%	10.00%	10.00%
20	-1.75%	4.64%	4.98%	5.00%	5.00%	5.00%	5.00%	5.00%	5.00%
30	1.76%	3.31%	3.33%	3.33%	3.33%	3.33%	3.33%	3.33%	3.33%
40	2.09%	2.50%	2.50%	2.50%	2.50%	2.50%	2.50%	2.50%	2.50%
50	1.89%	2.00%	2.00%	2.00%	2.00%	2.00%	2.00%	2.00%	2.00%
60	1.63%	1.67%	1.67%	1.67%	1.67%	1.67%	1.67%	1.67%	1.67%
70	1.42%	1.43%	1.43%	1.43%	1.43%	1.43%	1.43%	1.43%	1.43%
80	1.25%	1.25%	1.25%	1.25%	1.25%	1.25%	1.25%	1.25%	1.25%
90	1.11%	1.11%	1.11%	1.11%	1.11%	1.11%	1.11%	1.11%	1.11%
100	1.00%	1.00%	1.00%	1.00%	1.00%	1.00%	1.00%	1.00%	1.00%
200	0.50%	0.50%	0.50%	0.50%	0.50%	0.50%	0.50%	0.50%	0.50%
300	0.33%	0.33%	0.33%	0.33%	0.33%	0.33%	0.33%	0.33%	0.33%
400	0.25%	0.25%	0.25%	0.25%	0.25%	0.25%	0.25%	0.25%	0.25%
500	0.20%	0.20%	0.20%	0.20%	0.20%	0.20%	0.20%	0.20%	0.20%
600	0.17%	0.17%	0.17%	0.17%	0.17%	0.17%	0.17%	0.17%	0.17%
700	0.14%	0.14%	0.14%	0.14%	0.14%	0.14%	0.14%	0.14%	0.14%
800	0.13%	0.13%	0.13%	0.13%	0.13%	0.13%	0.13%	0.13%	0.13%
900	0.11%	0.11%	0.11%	0.11%	0.11%	0.11%	0.11%	0.11%	0.11%
1000	0.10%	0.10%	0.10%	0.10%	0.10%	0.10%	0.10%	0.10%	0.10%

Table C.2.: $d = f^{-1}(d', n)$

	10%	20%	30%	40%	50%	60%	70%	80%	90%
10	0.00%	20.87%	33.03%	44.39%	55.55%	66.67%	77.78%	88.89%	100.00%
20	9.78%	20.99%	31.58%	42.11%	52.63%	63.16%	73.68%	84.21%	94.74%
30	10.19%	20.69%	31.03%	41.38%	51.72%	62.07%	72.41%	82.76%	93.10%
40	10.22%	20.51%	30.77%	41.03%	51.28%	61.54%	71.79%	82.05%	92.31%
50	10.19%	20.41%	30.61%	40.82%	51.02%	61.22%	71.43%	81.63%	91.84%
60	10.17%	20.34%	30.51%	40.68%	50.85%	61.02%	71.19%	81.36%	91.53%
70	10.14%	20.29%	30.43%	40.58%	50.72%	60.87%	71.01%	81.16%	91.30%
80	10.13%	20.25%	30.38%	40.51%	50.63%	60.76%	70.89%	81.01%	91.14%
90	10.11%	20.22%	30.34%	40.45%	50.56%	60.67%	70.79%	80.90%	91.01%
100	10.10%	20.20%	30.30%	40.40%	50.51%	60.61%	70.71%	80.81%	90.91%
200	10.05%	20.10%	30.15%	40.20%	50.25%	60.30%	70.35%	80.40%	90.45%
300	10.03%	20.07%	30.10%	40.13%	50.17%	60.20%	70.23%	80.27%	90.30%
400	10.03%	20.05%	30.08%	40.10%	50.13%	60.15%	70.18%	80.20%	90.23%
500	10.02%	20.04%	30.06%	40.08%	50.10%	60.12%	70.14%	80.16%	90.18%
600	10.02%	20.03%	30.05%	40.07%	50.08%	60.10%	70.12%	80.13%	90.15%
700	10.01%	20.03%	30.04%	40.06%	50.07%	60.09%	70.10%	80.11%	90.13%
800	10.01%	20.03%	30.04%	40.05%	50.06%	60.08%	70.09%	80.10%	90.11%
900	10.01%	20.02%	30.03%	40.04%	50.06%	60.07%	70.08%	80.09%	90.10%
1000	10.01%	20.02%	30.03%	40.04%	50.05%	60.06%	70.07%	80.08%	90.09%

C. Experiments and Randomly Generated Instances

Table C.3.: Situations where ODSCP1 proofs infeasibility

d	100	200	300	400	500	600	700	800	900	1000	
10.0%	4	5	5	5	5	5	5	5	5	5	49
10.5%	5	5	5	5	2		3	5	4	4	38
11.0%	5	3		1			4	3	4	3	23
11.5%	5						2	2	1		10
12.0%	4					1					5
12.5%	2										2
13.0%	1										1
13.5%											0
	26	13	10	11	7	6	14	15	14	12	128

Table C.4.: Situations where ODSCP2 proofs infeasibility

d	100	200	300	400	500	600	700	800	900	1000	
10.0%	5	5	5	5	5	5	5	5	5	5	50
10.5%	5	5	5	5	3	1	3	2			29
11.0%	5	5		2	1						13
11.5%	5										5
12.0%	5										5
12.5%	3										3
13.0%	2										2
13.5%											0
	30	15	10	12	9	6	8	7	5	5	107

C.2. Experimenting on ODSCP1 and ODSCP2

The following tables report on an experiment that compares ODSCP1 and ODPC2 (see section 3), both optimizations versions of DSCP. The instances are generated by the algorithm described in section C.1, varying $n \in \{100, 200, \ldots, 900, 1000\}$ and $d \in \{10\%, 10.5\%, \ldots, 19.5\%, 20\%\}$. For each setting we gen C.3 and Table C.4 report the situations where CPLEX was able to proof infeasibility within one hour. If we compare the successful runs we can see that ODSCP1 performs better than ODSCP2. Looking a bit more closely at the tables we can investigate that ODSCP2 performs better for smaller instances, while OPCD2 performs much better in case of larger instances. The situation goes into the reverse when we compare the number of optimally solved instances. In total, we see that the performance is comparable and. ODSCP1 solves 459 (331 positive and 128 negative answers) of 1050 instances and ODSCP2 could find 454 (347 positive and 107 negative) answers.

Table C.5.: Situations where ODSCP1 gives the optimal solution

d	100	200	300	400	500	600	700	800	900	1000	
13.5%	1										1
14.0%	2										2
14.5%	4										4
15.0%	4	3	2								9
15.5%	5	5	3	2							15
16.0%	5	5	4	4	2	1					21
16.5%	5	4	5	3	5	5	3				30
17.0%	5	5	5	5	5	4	2				31
17.5%	5	5	5	5	5	5	5				35
18.0%	5	5	5	5	5	5	5				35
18.5%	5	5	5	5	5	5	5				35
19.0%	5	5	5	5	5	5	5	1	1		37
19.5%	5	5	5	5	5	5	5	2	1		38
20.0%	5	5	5	5	5	5	5	1	2		38
	61	52	49	44	42	40	35	4	4		331

Table C.6.: Situations where ODSCP2 gives the optimal solution

d	100	200	300	400	500	600	700	800	900	1000	
14.0%	2										2
14.5%	4	1									5
15.0%	4	4	2								10
15.5%	5	3	4	2							14
16.0%	5	5	5	2							17
16.5%	5	5	5	5	1	2					23
17.0%	5	5	5	5	4	3	3				30
17.5%	5	5	5	5	4	5	2	1	1		33
18.0%	5	5	5	5	5	5	5	1	1		37
18.5%	5	5	5	5	5	5	5	4	2	1	42
19.0%	5	5	5	5	5	5	5	4	4		43
19.5%	5	5	5	5	5	5	5	5	4		44
20.0%	5	5	5	5	5	5	5	5	5	2	47
	60	53	51	44	34	35	30	20	17	3	347

C. Experiments and Randomly Generated Instances

Table C.7.: Production costs

t	o_1^t	o_2^t
1	1	1
2	1	1
3	5	5

Table C.8.: For every combination of inventory levels we calculate the production quantities that minimize the expectation value of F

y_1^1	y_1^2	u_1^1	u_1^2	$\mathbf{E}(F)$	y_1^1	y_1^2	u_1^1	u_1^2	$\mathbf{E}(F)$	y_1^1	y_1^2	u_1^1	u_1^2	$\mathbf{E}(F)$
0	0	3	0	16	0	0	3	2	12	0	0	0	3	17
1	0	2	0	15	1	0	2	2	11	1	0	0	2	12
2	0	1	0	14	2	0	1	2	9.5	2	0	0	1	6.5
3	0	0	0	13	3	0	0	2	8.5	3	0	0	0	1.5
0	1	2	0	15	0	1	3	1	11	0	1	0	2	12
1	1	1	0	14	1	1	2	1	9.5	1	1	0	1	6.5
2	1	0	0	13	2	1	1	1	8.5	2	1	0	0	1.5
3	1	0	0	13	3	1	0	1	7.5	3	1	0	0	1.5
0	2	1	0	14	0	2	3	0	9.5	0	2	0	1	6.5
1	2	0	0	13	1	2	2	0	8.5	1	2	0	0	1.5
2	2	0	0	13	2	2	1	0	7.5	2	2	0	0	1.5
3	2	0	0	13	3	2	0	0	6.5	3	2	0	0	1.5
0	3	0	0	13	0	3	2	0	8.5	0	3	0	0	1.5
1	3	0	0	13	1	3	1	0	7.5	1	3	0	0	1.5
2	3	0	0	13	2	3	0	0	6.5	2	3	0	0	1.5
3	3	0	0	13	3	3	0	0	6	3	3	0	0	1.5

C.3. SDWLP: Instances and Solutions

For experiments with SPFLP we generated simple instances. The instance used in the experiments in section 2.3.1 we have to service 3 customers from 2 production sites within 3 periods. Each customer's demand equals one with certain probability. Transporting one unit costs 1 and also as well as the inventory. The capacity of the inventory is given by 3. The production costs are given in the Table C.7. The Output of the Dynamic Programing Procedure, as well as the heuristic is a table that gives the optimal production quantities to the current state of the inventory. In Table C.8 the solution is given for $p_i^t = 0.5$ for all $i \in \{1,2\}$ and $t \in \{1,2,3\}$.

List of Figures

1.1. Varignon frame 4
1.2. Example where 3 cameras are sufficient to cover the border, but 4 cameras are needed to cover the whole area 10
1.3. Example of 2 different triangulations 10
1.4. An example where putting the camera on the wall leads to a better solution . . 11
1.5. An example where putting the camera inside the polygon leads to a better solution 11

2.1. Sequencing of decisions 15
2.2. Comparison of different solutions 19
2.3. Distributions of the optimal solution to instances with different levels of probability $(p_j^{(t)} = p \in \{0, 0.01, 0.02, \ldots, 1\})$ 22
2.4. Choice of sample size N and the number of samples M 22
2.5. Expected costs to different levels of inventory y $(1 : [0,0]; 2 : [0,1]; \ldots; 4 : [0,3]; 5 : [1,0]; \ldots; 12 : [2,3])$. 22

3.1. Example for a road network 26
3.2. Polygon with two independent surveillance systems 27
3.3. Example of a transformation of $\mathcal{F} = \{(\{2\}, \{2\}), (\{1, 3\}, \{1\}), (\{1, 2\}, \{1, 2\}), (\{2, 3\}, \{1, 2\}), (\{1\}, \{3\}), (\{3\}, \{1, 2\})\}$ 32
3.4. Transformation of literal i and clause j into circular configurations 37
3.5. Example of a clause c_j in tree variables $x_1\ x_2\ x_3$ 38
3.6. Example where the clause c_j=false 38
3.7. Example where the clause c_j=true 39
3.8. Example of a transformation of a 3-SAT instance with 3 clauses and 3 variables 46
3.9. Solution to the example given in Figure 3.8 47
3.10. Example of a transformation of PPMP to DSCP to apply Proposition 3.3.15 . . 47
3.11. Complexity of DSCP 50
3.12. Example of a transformation from Max-2-SAT to ODSCP2 58
3.13. Complexity of ODSCP 60

List of Figures

3.14. The derived graph for the following SET COVER instance: $S = \{v_1, v_2, v_3, v_4, v_5, v_6\}$ and $\mathcal{C} = \{S_1, S_2, S_3, S_4\}$ where $S_1 = \{v_1, v_3, v_6\}, S_2 = \{v_2, v_4, v_5\}, S_3 = \{v_3, v_4, v_6\}$ and $S_4 = \{v_3, v_4, v_6\}$. 69
3.15. Ants when choosing the shortest way between their nest and a food source . . . 71
3.16. Methodology 75
3.17. CPLEX solution process for both [*GRNP-MIP*] and [*RMIP-GRNP*], on instance 100-21-4.dat 76
3.18. A small cut-out of the resulting network 80
3.19. Junction types 84
3.20. Cubic graph 84
3.21. Polygon 85
3.22. $C_{3,3}$ a non-planar graph 86
3.23. Junction of corridors 86
3.24. Junction in detail 86
3.25. Transformation of 3-SAT to vertex guard [81] 88
3.26. Representation of a clause: the white nodes represent the values of the literals . 88
3.27. Narrow corridor with an alcove and spike 89
3.28. Transformation of 3-SAT to the Vertex Guard Double Cover Problem 90
3.29. (Chvátal's Comb): an example where the bound is sharp 91
3.30. A polygon with one hole gets transformed into a polygon without holes to select the guard sets 91
3.31. Gluing a cut: we have 2 possibilities to choose the second color 92
3.32. Polygon with three neighboring holes can be transformed to a polygons with one hole 93
3.33. Catalan numbers 94
3.34. Comparing the optimal solution with the best solution reachable by the triangulation technique 94
3.35. The optimal solution uses two vertices, while the number of vertices needed by solutions that correspond to triangulations is dependent on the number of spikes 95

Bibliography

[1] M. Andrec, B. N. Kholodenko, R. M. Levy, E. Sontag, Inference of signaling and gene regulatory networks by steady-state perturbation experiments: structure and accuracy. J. Theoret. Biol. 232 (2005) 427–441

[2] M.I. Arnone, E.H. Davidson, The hardwiring of development: organization and function of genomic regulatory systems. Development 124 (1997) 1851–1864

[3] M. Bansal, G. D. Gatta, D. di Bernardo, Inference of gene regulatory networks and compound mode of action from time course gene expression profiles. Bioinformatics 22 (2006) 815–822

[4] D.P. Bertsekas: Dynamic Programming and Optimal Control, Athena Scientific, Belmont, MA (1995)

[5] M. Benkert,t. Shirabe, A. Wolff: The Minimum Manhattan Network Problem - Approximations and Exact Solutions, citeseer.ist.psu.edu/benkert04minimum.html

[6] M. Benkert, F. Widmann, A. Wolff: The Minimum Manhattan Network Problem - A Fast Factor-3 Approximation, citeseer.ist.psu.edu/702298.html

[7] B. Bozkaya, J. Zhang, E. Erkut: An Efficient Genetic Algorithm for the p-Median Problem, in Z. Drezner and H.W. Hamacher (Eds.) Facility Location: Applications and Theory, Springer, New York (2002) 179-205

[8] J. Bower, H. Bolouri, Computational Modeling of Genetic and Biochemical Networks. The MIT Press, Cambridge, MA (2001)

[9] M. P. Brynildsen, L. M. Tran, J. C. Liao, A Gibbs sampler for the identification of gene expression and network connectivity consistency. Bioinformatics 22 (2006) 3040–3046

[10] P. Brazhnik, Inferring gene networks from steady-state response to single-gene perturbations. J. Theoret. Biol. 237 (2005) 427–440

Bibliography

[11] T. Chen, H. He, G. Church, Modeling gene expression with differential equations, Pac. Symp. Biocomput. 4 (1999) 29-40

[12] T. Chen, V. Filkov, S.S. Skiena, Identifying gene regulatory networks from experimental data. Parallel Computing 27 (2001) 141–162

[13] J. Cheriyan, R. Ravi: Lecture Notes on Approximation Algorithms for Network Problems: Media, url = "http://www.math.uwaterloo.ca/ jcheriya/lecnotes.html"

[14] X. Chen, G. Anantha, X. Wang An effective structure learning method for constructing gene networks. Bioinformatics 22/11 (2006) 1367–1374

[15] R. Cho, M. Campbell, E. Winzeler, L. Steinmetz, A. Conway, L. Wodicka, T. Wolfsberg, A. Gabrielan, D. Landsman, D. Lockhart, R. Davis, A genome-wide transcriptional analysis of the mitotic cell cycle. Mol. Cell 2 (1998) 65-73

[16] S. Chopra, C.-Y. Tsai: Polyhedral Approaches for the Steiner Tree Problem on Graphs, in D.-Z. Du and X. Cheng (Ed.), Steiner Trees in Industries

[17] H. S. M. Coxeter: Introduction to Geometry John Wiley & Sons, New York - London (1961)

[18] Steven A. Cook: The Complexity of Theorem-Proving Procedures: Annual ACM Symposium on Theory of Computing, Shaker Heights, Ohio, United States,(1971) 151–158

[19] E. S. Correa, M. T. A. Steiner, A. A. Freitas, C. Carnieri: A Genetic Algorithm for the P-median Problem, in LE Spector and E Goodman et al. (Eds.): Proc. 2001 Genetic and Evolutionary Computation Conference (GECCO-2001), Morgan Kaufmann, San Fracisco (2001) 1268-1275

[20] M.C. Couto, C.C. de Souza, P.J. de Rezende: An Exact and Efficient Algorithm for the Orthogonal Art Gallery Problem, Brazilian Symposium on Computer Graphics and Image Processing (2007) 87–94

[21] S. D'Agostino, Personal Communication (2007).

[22] M. De Hoon, S. Imoto, S. Miyano, Inferring gene regulatory networks from time-ordered gene expression data using differential equations. Lecture Notes in Computer Science 2534 (2002) 267–274

[23] H. De Jong, Modeling and Simulation of Genetic Regulatory Systems: A Literature Review. J. of Computational Biology 9/1 (2002) 67–103

Bibliography

[24] D. Di Bernardo, T.S. Gardner, J.J. Collins, Robust Identification of Large Genetic Networks. Pacific Symposium on Biocomputing 9 (2004) 486–497

[25] R. Diestel: Graph Theory, Springer-Verlag, Heidelberg, New York (2005)

[26] K. Doerner, R.F. Hartl, M. Karall, M. Reimann: Heuristic solution of an extended double-coverage ambulance location problem for Austria, Central European J. of Operations Research 13 (2005) 325–340

[27] W. Domschke, A. Drexl: Logistik, Bd.3, Standorte, Oldenbourg Wiss., Mchn. (1996)

[28] M. Dorigo, V. Maniezzo, A. Colorni, The Ant System: Optimization by a colony of cooperating agents. IEEE Transactions on Systems, Man, and Cybernetics Part B: Cybernetics 26/1 (1996) 29–41

[29] M. Dorigo, T. Stützle, The Ant Colony Optimization Metaheuristic: Algorithms, Applications, and Advances. In F. Glover, G. Kochenberger (Eds.), Handbook of Metaheuristics (2002) 251–285

[30] M. Dorigo, T. Stützle, Ant Colony Optimization. MIT Press, Cambridge, MA (2004)

[31] Z. Drezner: Facility Location a Survey of Applications and Methods (Springer Series In Operations Research), Springer Verlag, New York, Berlin Heidelberg (1995)

[32] S. Eidenbenz: Inapproximability Results for Guarding Polygons without Holes, ISAAC: 9th International Symposium on Algorithms and Computation, Organized by SIGAL of the IPSJ and IEICE (1998)

[33] D. Erlenkotter: A Dual Based Procedure for Uncapacitated Facility Location, Operations Research, 26 (1978) 992-1009

[34] P.L. Hammer: Annals of Operations Research: Recent Developments in the Theory and Applications of Location Models 1/2, Kluwer Academic Publishers (2002)

[35] U. Feige: A treshhold of $\ln(n)$ for approximating set cover, Proc. 28 Annual ACM Symp. on Theory of Computing (1996) 314–318

[36] T.A. Feo, M.G.C. Resende, A probabilistic heuristic for a computationally difficult set covering problem. Operations Research Letters 8 (1989) 67-71

[37] T.A. Feo, M.G.C. Resende, Greedy randomized adaptive search procedures. J. of Global Optimization 6 (1995) 109-133

Bibliography

[38] V. Filkov, S. Skiena, J. Zhi, Analysis Techniques for Microarray Time-Series Data. J. of Comput. Biol. 9/2 (2002) 317–330

[39] V. Filkov, Identifying Gene Regulatory Networks from Gene Expression Data. In S. Aluru (Ed.), Handbook of Computational Molecular Biology, Chapman&Hall/CRC Press (2005)

[40] S. Fisk: A short proof of Chvátal's watchman theorem. Journal of Combinatorial Theory Ser B 24, (1978) 374

[41] R. Francis, L. Richard, L. F. McGinnis, J. A. White: Facility Layout and Location, 2nd. ed., Prentice Hall, Englewood Cliffs, NJ (1992)

[42] N. Friedman, M. Linia, I. Nachman, D. Peer, Using Bayesian networks to analyze expression data. J. of Comput. Biol. 7 (2000) 601-620

[43] N. Friedman, Inferring Cellular Networks Using Probabilistic Graphical Models. Science 303 (2004) 799–805

[44] M.R. Garey, R.L. Graham, D.S.Johnson: The complexity of computing Steiner minimal trees. SIAM J. of Appl. Math. 31 (1977) 835–859

[45] M. R. Garey, D. S. Johnson, The Rectilinear Steiner Tree Problem is NP-Complete, SIAM Journal on Applied Mathematics 32/4 (1977) 826–834

[46] M.R. Garey, D.S. Johnson, Computers and Intractability: A Guide to the Theory of NP-completeness. W.H. Freeman and Company, New York (1979)

[47] A.L. Gartel, S.K. Radhakrishnan, Lost in Transcription: p21 Repression, Mechanisms, and Consequences. Cancer Res. 65/10 (2005) 3980–3985

[48] J. Gebert, N. Radde, G.W. Weber, Modeling gene regulatory networks with piecewise linear differential equations. European Journal of Operational Research 181/3 (2007) 1148–1165

[49] M. Gendreau, G. Laporte, F. Semet: Solving an ambulance location problem by tabu search, Location Science 5 (1997) 75-88

[50] F. Glover, M. Laguna, R. Mart: Fundamentals of Scatter Search and Path Relinking, Control and Cybernetics 29/3 (2000) 653-684

[51] G. Ghiani, G. Laporte, R. Musmanno: Introduction to Logistics Systems Planning and Control, Halsted Press, New York, NY, USA (2004)

[52] A. J. Goldman: Optimal center location in simple networks. Transp. Sci., 5 (1971) 240–255

[53] S . K . Ghosh: Approximation algorithms for art gallery problems Proc. Canadian Inform. Process. Soc. Congress (1987)

[54] S. Goss, S. Aron, J.L. Deneubourg, J.M. Pasteels, Self-organized shortcuts in the Argentine ant. Naturwissenschaften 76 (1989) 579–581

[55] J. Gudmundsson, C. Levcopoulos, G. Narasimhan: Approximating a Minimum Manhattan Network, Nordic Journal of Computing 8/2 (2001) 216-229

[56] S. L. Hakimi: Optimum Locations of Switching Centers and the Absolute Centers and Medians of a Graph Operations Research 12/3 (1964) 450–459

[57] I. Heller, C.B.Tompkins, An Extension of a Theorem of Dantzig's, in H.W. Kuhn, A.W. Tucker, Linear Inequalities and Related Systems, vol. 38, Annals of Mathematics Studies, Princeton (NJ): Princeton University Press (1956) 247–254

[58] D.S. Hochbaum, N. Megiddo, J. Naor, and A. Tamir: Tight bounds and 2approximation algorithms for integer programs with two variables per inequality. Mathematical Programming 62 (1993) 69–83

[59] A. Hoffman and J. Kruskal: Integral boundary points of convex polyhedra, in H. Kuhn, A. Tucker (eds.): Linear Inequalities and Related Systems, Princeton University Press (1956) 223-246

[60] K. Hogan, C. Revelle: Concepts and Applications of Backup Coverage, Management Science, Vol. 32, No. 11 (1986) 1434–1444

[61] I. Holyer: The NP-completeness of edge-coloring. SIAM J. Comput. 10 (1981) 718-720

[62] T. Homem-de-Mello, Variable-Sample Methods for Stochastic Optimimization, ACM Trans. on Modeling and Computer Simulation (2003)

[63] Mathematical Gems II, Ed: R. Honsberger, Washington DC (1976)

[64] S. Imoto, T. Higuchi, T. Goto, S. Miyano, Error tolerant model for incorporating biological knowledge with expression data in estimating gene networks. Stat. Methodol. 3/1 (2006) 1–16

[65] P. Jaillet: Probabilistic Travelling Salesman Problems, Ph.D. thesis, MIT (1985)

Bibliography

[66] R. Karp: Reducibilities among combinatorial problems, in R.E. Miller, J.W.Thatcher (Eds.) : Complexity of Computer Computations, Plenum Press (1972) 85-103

[67] S. Kauffman: Metabolic stability and epigenesis in randomly constructed genetic nets. J. Theor. Biol. 22 (1969) 437-467

[68] B.N. Kholodenko, A. Kiyatkin, F. Bruggeman, E.D. Sontag, H. Westerhoff, J. Hoek: Untangling the wires: a novel strategy to trace functional interactions in signaling and gene networks. PNAS 99 (2002) 12841–12846

[69] P.M. Kim, B. Tidor, Limitations of Quantitative Gene Regulation Models: A Case Study. Genome Res. 13 (2003) 2396–2405

[70] T. Koch, A. Martin: Solving Steiner tree problems in graphs to optimality Networks 32/3 (1998) 207–232

[71] A. Kleywegt, A. Shapiro, T. Hohem-de-Mello: The Sample Average Method for Stochastic Discrete Optimization, SIAM J. Optim. 12 (2001/02)

[72] L. Lania, B. Majello, G. Napolitano, Transcriptional Control by Cell-Cycle Regulators: A Review. J. of Cellular Physiology 179 (1999) 134-141

[73] G. Laporte, P. J. Dejax: Dynamic Location-Routeing Problems, J. Opl Res. Soc. 40/5 (1989) 471–482

[74] G. Laporte, F. Louveaux, L. Van Hamme: Exact Solution to a Location Problem with Stochatic Demands, Transportation Science 28 (1994)

[75] G. Laporte, F.Louveaux, H. Mercure: Models and exact solutions for a class of stochastic location-routing problems. European Journal of Operational Research 39 (1989) 7178

[76] C. Lax, S. Fogel, C. Cramer: Regulatory mutants at the his1 locus of yeast. Genetics 92/2 (1979) 363–82

[77] R. Laubenbacher, B. Stigler, A computational algebra approach to the reverse engineering of gene regulatory networks. J. Theoret. Biol. 229/4 (2004) 523–537

[78] N. H. Lee: Genomic approaches for reconstructing gene networks. Pharmacogenomics 6/3 (2005) 245–258

[79] Y. Li: A Newton Acceleration of the Weiszfeld Algorithm for Minimizing the Sum of Euclidean Distances, Computational Optimization and Applications, Volume 10/3 (1998) 219-242

Bibliography

[80] D. Lichtenstein: Planar formulae and their uses, SIAM Journal on Computing 11 (1982) 329–343

[81] D.T. Lee, A.K. Lin: Computational Complexity of Art Gallery Problems. IEEE Transactions on Information Theory, 32-2 (1986) 276–282

[82] G. Lulli, M. Romauch: Inferring gene regulatory networks by mathematical programming. to appear in Discrete Applied Mathematics

[83] A. Makhorin:GNU Linear Programming Kit - Reference Manual - Version 4.1, Department for Applied Informatics, Moscow Aviation Institute, Moscow, Russia (2003)

[84] J. MacGregor Smith: Steiner Minimal Trees in E^3: Theory, Algorithms, and Applications, url = "citeseer.ist.psu.edu/486856.html"

[85] A.A. Margolin, I. Nemenman, K. Basso, C. Wiggins, G. Stolovitzky, R. Dalla Favera, A. Califano, ARACNE: an algorithm for the reconstruction of gene regulatory networks in a mammalian cellular context. BMC Bioinformatics 7 (2006) Suppl 1:S7

[86] M.T. Melo, S. Nickel, F. Saldanha da Gama: Large-Scale Models for Dynamic Multi-Commodity Capacitated Facility Location, Berichtsreihe des Fraunhofer Inststituts für Techno- und Wirtschaftsmathematik (ITWM) Kaiserslautern (2003)

[87] William Miehle: Link-Length Minimization in Networks, Operations Research 6/2 (1958) 232–243

[88] H. Min, V. Jayaraman, R. Srivastava: Theory and Methodology, Combined location-routing problems: A synthesis and future research directions, European Journal of Operational Research 108 (98) 1–15.

[89] A. Murray, K. Kim, J. Davis, R. Machiraju, R. Parent: Coverage Optimization to Support Security Monitoring, Computers, Environment and Urban Systems 31 (2007) 133-147.

[90] G. Nagy, S. Salhi: Location-routing: Issues, models and methods, European Journal of Operational Research 177 (2007) 649672

[91] S. Hesse Owen, M.S. Daskin: Strategic Facility Location: A Review, Eropean Journal of Operational Research 111 (1998)

[92] J. O'Rourke: Art gallery theorems and algorithms. Oxford University Press (1987)

Bibliography

[93] B.O. Palsson, In silico biotechnology: Era of reconstruction and interrogation. Current Opinion in Biotechnology, 15/1 (2004) 50–51.

[94] D. Peer, A. Regev, G. Elidan, N. Fridman, Inferring subnetworks from expression profiles. Bioinformatics 17 (2001) 215-224.

[95] B. Perrin, L. Ralaivola, A. Mazurie, S. Bottani, J. Mallet, F. d' Alch Buc, Gene networks inference using dynamic Bayesian networks. Bioinformatics 19 (2003) 138–148.

[96] A. J. Pittard, B.E. Davidson, TyrR protein of Escherichia coli and its role as repressor and activator. Molecular microbiology, 5/7 (1991) 1585–1592

[97] A. Remenyi, M.C. Good, R.P. Bhattacharyya, W.A. Lim, The role of docking interactions in mediating signaling input, output, and discrimination in the yeast MAPK network. Mol Cell. 22;20/6 (2005) 951–962.

[98] M. G. C. Resende: Computing Approximate Solutions of the Maximum Covering Problem with GRASP, Journal of Heuristic 4/2 (1998) 161–177.

[99] M.G.C. Resende, C.C. Ribeiro, Greedy randomized adaptive search procedures. In F. Glover, G. Kochenberger (Eds.), Handbook of Metaheuristics (2002) 219–249.

[100] R. Rizzi, Personal Communication (2007).

[101] M. Romauch, R.F. Hartl: Dynamic Facility Location with Stochastic Demands. SAGA 2005: 180–189

[102] J. B. Rosen, G. L. Xue: On the Convergence of Miehle's Algorithm for the Euclidean Multifacility Location Problem (in Technical Note), Operations Research, Vol. 40/1 (1992) 188–191

[103] E. Sakamoto, H. Iba, Inferring a system of differential equations for a gene regulatory network by using genetic programming. Proceedings of the Congress on Evolutionary Computation (2001) 720–726

[104] S. Salhi, G. Nagy: Local improvement in planar facility location using vehicle routing facility location using vehicle routing. Annals of Operations Research (2007)

[105] T. Santoso, S. Ahmed, M. Goetschalckx, J. Shapiro: A Stochastic Programming Approach for Supply Chain Network Design under Uncertainty, The Stochastic Programming E-Print Series (SPEPS), (2003)

[106] K.A. Schafer, The Cell Cycle: A Review. Vet. Pathol. 35 (1998) 461–478

[107] D. Schilling, V. Jayaraman, R. Barkhi: A review of covering problems in facility location. Location Science, 1 (1993) 25-55

[108] C. Schneeweiss: Dynamisches Programmieren. Physica, Heidelberg (1974)

[109] D. Schuchardt, H.-D. Hecker: Two NP-Hard Art-Gallery Problems for Ortho-Polygons. Math. Log. Q. 41 (1995) 261-267

[110] P. Sebastiani, E. Gussoni, I.S. Kohane, M.F. Ramoni, Statistical Challenges in Fuctional Genomics. J. of Statistical Science 18/1 (2003) 33–70

[111] H.D. Sherali, W.P. Adams, A tight linearization and an algorithm for 0-1 quadratic programming problems. Management Science 32/10 (1986) 1274–1290

[112] T. Shermer: Recent results in art galleries, Proc. IEEE, 80 (1992) 1384-1399

[113] Y. Shi, T. Mitchell, Z. Bar-Joseph, Inferring Gene Regulatory Relationships from Multiple Time Series Datasets. Bioinformatics 23/6 (2007) 755–763

[114] M. Shimazu, T. Sekito, K. Akiyama, Y. Ohsumi, Y. Kakinuma, A Family of Basic Amino Acid Transporters of the Vacuolar Membrane from Saccharomyces cerevisiae. J. Biol. Chem. 280/6 (2005) 4851–4857

[115] I. Shmulevich, E.R. Dougherty, S. Kim, W. Zhang, Probabilistic boolean networks: A rule-based uncertainty model for gene regulatory networks. Bioinformatics 18 (2002) 261–274

[116] L. Snyder: Facility location under uncertainty: a review, IIE Transactions 38/7 (2006) 547–564

[117] P. Spellman, G. Sherlock, M. Zhang, V. Iyer, K. Anders, M. Eisen, P. Brown, D. Botstein, B. Futcher, Comprehensive identification of cell cycle-regulated genes of the yeast Saccharomyces cerevisiae by microarray hybridization. Mol. Biol. Cell 9 (1998) 3273-3297

[118] R. P. Stanley, Enumerative combinatorics, Wadsworth Publ. Co., Belmont, CA, USA (1986)

[119] D. Thieffry, R. Thomas, Qualitative Analysis of Gene Networks. Pacific Symposium on Biocomputing 3 (1998) 77–88

Bibliography

[120] H.K. Tsai, H.H.S. Lu, W.H. Li, Statistical methods for identifying yeast cell cycle transcription factors. PNAS 102/38 (2005) 13532-13537

[121] J.Urrutia: Art Gallery and Illumination Problems, In J. Sac, J. Urrutia (Eds.), Handbook on Computational Geometry, Elsevier Science Publishers, Amsterdam (2000) 973–1027

[122] E.P. van Someren, L.F.A. Wessels, E. Backer, M.J.T. Reinders, Genetic network modeling. Pharmacogenomics 3/4 (2002) 507–525

[123] T.T. Vu, J. Vohradsky, Nonlinear differential equation model for quantification of transcriptional regulation applied to microarray data of Saccharomyces cerevisiae. Nucleic Acids Res. 35 (2007) 279–287

[124] D. M. Warme, P. Winter, and M. Zachariasen: Exact Algorithms for Plane Steiner Tree Problems: A Computational Study. In D.-Z. Du, J. M. Smith, and J. H. Rubinstein (ed.), Advances in Steiner Trees, pp. , Kluwer Academic Publishers, Boston, (2000) 81-116

[125] E. Weiszfeld: Sur le point pour lequel la somme des distances de n points dennés est minimum, Tôhoko Mathematics Journal, 43 (1937) 355–386

[126] C.J. Zhang, M.M. Cavenagh, R.A. Kahn, A family of Arf effectors defined as suppressors of the loss of Arf function in the yeast Saccharomyces cerevisiae. J Biol Chem. 31; 273/31 (1998) 19792–19796

[127] W. Zhao, E. Serpedin, E.R. Dougherty, Inferring gene regulatory networks from time series data using the minimum description length principle. Bioinformatics 22/17 (2006) 2129–2135

I want morebooks!

Buy your books fast and straightforward online - at one of the world's fastest growing online book stores! Environmentally sound due to Print-on-Demand technologies.

Buy your books online at

www.get-morebooks.com

Kaufen Sie Ihre Bücher schnell und unkompliziert online – auf einer der am schnellsten wachsenden Buchhandelsplattformen weltweit!
Dank Print-On-Demand umwelt- und ressourcenschonend produziert.

Bücher schneller online kaufen

www.morebooks.de

OmniScriptum Marketing DEU GmbH
Heinrich-Böcking-Str. 6-8
D - 66121 Saarbrücken
Telefax: +49 681 93 81 567-9

info@omniscriptum.com
www.omniscriptum.com

Printed by Books on Demand GmbH, Norderstedt / Germany